*Other books by Randy A. Steinberg:*
Implementing ITIL®
Adapting Your IT Organization to the Coming Revolution in IT Service Management
Trafford Publishing: ISBN 1-4120-6618-2



Measuring ITIL® Measuring, Reporting and Modeling the IT Service
Management Metrics That Matters Most To IT Senior Executives



Cover design by Anthony Mingo, Design & Art Direction



Note for Librarians: A cataloguing record for this book is available from Library and Archives Canada at
www.collectionscanada.ca/amicus/index-e.html
ISBN 1-4120-9392-9

*Printed in Victoria, BC, Canada. Printed on paper with minimum 30% recycled fibre.*
*Trafford's print shop runs on "green energy" from solar, wind and other environmentally-friendly power sources.*

*Offices in Canada, USA, Ireland and UK*

**Book sales for North America and international:**
Trafford Publishing, 6E–2333 Government St.,
Victoria, BC V8T 4P4 CANADA
phone 250 383 6864 (toll-free 1 888 232 4444)
fax 250 383 6804; email to orders@trafford.com
**Book sales in Europe:**
Trafford Publishing (UK) Limited, 9 Park End Street, 2nd Floor
Oxford, UK OX1 1HH UNITED KINGDOM
phone 44 (0)1865 722 113 (local rate 0845 230 9601)
facsimile 44 (0)1865 722 868; info.uk@trafford.com
**Order online at:**
trafford.com/06-1146

10 9 8 7 6 5 4 3 2

# Measuring ITIL®

## Measuring, Reporting and Modeling the IT Service Management Metrics That Matter Most To IT Senior Executives

**Randy A. Steinberg**

# Dedication

This book is dedicated to those very hard working IT professionals and managers who deserve to see their IT solutions deploy and operate day-to-day at acceptable cost to their company.

***IT faces a serious challenge…***

- ✓ It is the only business organization that almost never measures its operational effectiveness and efficiency
- ✓ It seldom measures the costs incurred for the services it delivers outside the budget it is given
- ✓ It monitors technologies but almost never monitors labor in terms of rework, waste and misuse
- ✓ It implements technologies with little measurement of deficiencies and defect rates

In short, IT rarely monitors and manages to the metrics that matter most to IT Senior Executives. Worse yet, the following basic concepts seem to bypass many IT management organizations concerning the services they support and deliver:

*"If you don't measure it, you can't manage it"*

*"If you don't measure it, you can't improve it"*

*"If you don't measure it, you probably don't care'"*

*"If you can't influence it, then don't measure it"*

IT can no longer continue to operate this way.
It's time to operate IT like a Service Organization.

*If you disagree, kindly leave this book for someone else to read.*

- The Author

# Table of Contents

# Chapter 1

# Book Overview

## *1.1 Why This Book Was Written*

This book is written in the hopes of helping IT to overcome the metrics gap that exists in many of today's IT organizations. Much has been written lately about how IT needs to be "run like a business". A stronger statement is that IT needs to "act like a business". While technology is certainly important, IT needs to inject much better business management practices by operating services instead of technology silos, measuring the quality and effectiveness of those services and taking timely actions to make sure those services are delivered in line with business needs.

This book focuses on the metrics reporting aspects of running an IT services operation. It assumes that the reader is familiar with ITIL concepts – maybe even in the throes of implementing ITIL practices. Without appropriate metrics, there is no way that those efforts can be validated and their benefits quantified. Without "metrics that matter", it is impossible for any ITIL effort to ever demonstrate its value or operate in a cycle of continuous improvement.

In working with many companies around the globe, it is surprising how many IT professionals really have few ideas on what should actually be measured or how an ITSM metrics program should be run. That is where this book steps in. It attempts to present, in practical terms, how an ITSM Metrics Program can be put together and provides suggested sets of metrics that can be used for every ITIL process.

A recent COMPASS published survey discovered some very dismal facts about IT and how it is measuring its services. When asked how well business organizations actually measured their IT processes, only 4 percent of respondents felt able to say that all of their ITIL processes were fully measured for maturity and less than one third of respondents had maturity measures for all ITIL processes. About 55% cite "some" measures, 13% said they had "absolutely no measures at all". These kinds of results would be totally unacceptable in any other business organization, yet IT has been operating this way for quite some time.

The metrics in this book are suggested recommendations meant as a starting point for any ITSM Metrics Program. It is fully expected that readers will take these and customize them to what fits into their organizations based on unique needs, communication styles and how things are reported. There book attempts to stay as practical as possible. A keen focus was on providing solutions that any organization can implement "starting on Monday morning" versus a lot of metrics theory.

## *1.2 Contents of the CD Included With This Book*

Included with this book is a CD that contains two simple tools that you may find helpful:

**DICE Model Tool –** *Dice Model.xls*

Chapter 17 describes this tool in more detail. The tool itself can be used to predict the success outcome of any ITSM initiative or improvement effort. It runs as a Microsoft EXCEL® spreadsheet.

**ITSM Metrics Modeling Tool –** *ITSM Metrics Model.xls*

Chapter 18 describes this tool in more detail. The tool itself can be used to measure, report and model the ITSM Metrics that matter. It utilizes many of the concepts discussed throughout this book. It runs as a Microsoft EXCEL® spreadsheet.

In addition to the above, the CD also includes electronic versions of Chapters 17 and 18 in Acrobat PDF® format. These are stored as:

*Using the Dice Model.pdf*

*Using the ITSM Metrics Tool.pdf*

## *1.3 Book Chapters in Brief*

Brief descriptions of remaining book chapters are as follows:

**Chapter 2 – An Overview of ITSM Metrics**

This chapter presents a high level overview of the need for metrics from an ITSM perspective. It covers the difference between typical IT measurements and what metrics matter the most to senior executives.

**Chapter 3 – The ITSM Metrics Model**

This chapter covers basics of a suggested metrics model that can be used with ITSM. It covers basic concepts such as Operational Metrics, Key Performance Indicators (KPIs), Critical Success Factors (CSFs), and other outputs of an ITSM Metrics Program.

**Chapters 4 to 14 – ITIL Metrics Chapters**

These chapters list suggested metrics for each ITIL process. Each chapter covers one ITIL process such as Incident Management, Problem Management, Change Management, and so on. For each ITIL process, a set of operational metrics, KPIs and CSFs is given. Suggested calculations for translating Operational Metrics into KPIs and CSFs are also shown.

**Chapter 15 – IT Workforce Management Metrics**

While Workforce Management is not explicitly covered within the current ITIL literature, this measurement of key workforce items such as staff turnover, skill levels, labor waste (non-value labor) cannot be ignored. This chapter covers suggested operational metrics, KPIs and CSFs for this area.

### Chapter 16 – Alternatives If Few Metrics Available

This chapter covers some ideas, tips and approaches that can be used if it is discovered that tools and capabilities are lacking for capturing many IT metrics.

### Chapter 17 – Using the DICE Model Tool

On the CD that accompanies this book, a copy of the DICE Model Tool is included. This chapter discusses how the tool can be used to predict the success outcome of your ITSM improvement projects and initiatives.

### Chapter 18 – Using the ITSM Metrics Modeling Tool

On the CD that accompanies this book, a copy of the ITSM Metrics Modeling Tool is included. This chapter discusses how the tool can be used to measure, report and model ITSM metrics that matter.

### Chapter 19 – Implementing an ITSM Metrics Program

This chapter presents an end state picture of how an ITSM Metrics program might operate. It also presents a generalized approach for how such a program might be implemented.

# Chapter 2

# An Overview of ITSM Metrics

## *2.1 The Importance of ITSM Metrics*

IT hates metrics.

IT is about the only business function that that rarely measures itself. Take a look at a manufacturing shop floor. Don't they carefully monitor labor, waste, defects, or efficiency? How about the corporate office? Is there not careful attention to things like stock price, earnings per share, revenue per headcount?

Yet, for some reason, IT has come along these many decades and has little in the way of best practices around measuring what they deliver. This has led to many problems and issues such as:

- Throwing money at ITSM projects with the goal of putting a process in place versus solving a business problem (Maybe thinking problems will go away if only that process was in place – a very risky assumption…)

- Inability to justify ITSM initiatives with business management and stakeholders (Management has little respect for initiatives when benefits cannot be seen or measured)

- Poor management decision making (hard to make the right decisions when you are blind to what is going on within your IT operational practices)

A number of IT people may be reading this and thinking something like: "What is this about? We do tons of metric reporting – even issue a major monthly report to management!" This may be so, but the real question here is this:

- Does anyone really read through the report?

- Can management make timely and accurate business decisions based on its content?

- Are the metrics chosen helpful to making decisions or are they just relating historical events that took place?

In short:

*Are these the metrics that really matter?*

## 2.2 *What Are Metrics That Matter?*

Having spent much time in the IT organizations of numerous companies, one will find that many IT shops are perfectly comfortable in telling you things like:

- How many IT changes were implemented
- How many incidents of some type occurred
- Current peak utilization of components such as network lines or servers
- How available an application or system was

These are certainly important to know. For now, start to keep them in mind as *Operational Metrics*. However, here are examples of a few painful questions to ask about how well the IT Service Delivery and Support practices really operate:

- How much operational labor is waste versus providing value to the business organization?
- What is your efficiency and effectiveness rate for processing IT changes?
- What is the labor utilization incurred for reacting to incidents and problems?
- What is the defect rate on releases put into production?
- What is the customer impact rate for incidents and problems?

From a management point of view, true *KPIs*, or *Key Performance Indicators*, provide a basis for making business decisions. The previous items are examples of indicators that require a management decision. Some examples using the above:

> Poor efficiency and effectiveness rates may indicate action is needed to reduce wasted labor when changes are being handled and processed
>
> Additional operational staff may be needed if there will not be enough labor to handle the change workload created by a new application or impending merger
>
> Incidents and problem rates may be high, but customer impact is low – therefore, IT is doing a fantastic job of protecting services but may not be able to sustain this if business volumes start to increase

In order to get to these decisions, yet another type of metric is needed to indicate when to take actions. This kind of metric is referred to as a *Tolerance*. This is an indicator that identifies, in advance, the boundary in which your company expects a KPI to operate and behave.

Organizations may elect to use one or more *Tolerance* metrics to bound KPI results for action decisions. If we use the Customer Impact Rate for Incidents KPI (as an example), an organization may elect to:

- ✓ Take no action if the KPI scores an impact rate of less than 10 percent
- ✓ Prioritize problem trending activities if the KPI scores an impact rate between 10 and 15 percent
- ✓ Establish an availability SWAT team and provide possible customer benefits and incentives if the KPI results score over 15 percent

In the above example, there are two Tolerance metrics in place that correspond to the Customer Impact Rate KPI. These are 10% and 15%. They represent the upper and lower bounds by which actions will be taken if the KPI falls inside or above those bounds.

Hopefully, at this point, you can now begin to see the difference between reporting on a historical operational event (such as "n incidents that took place") versus a metric that indicates a decision or action needs to occur.

## *2.3 Resistance to Using IT Metrics*

It is not clear how or why IT has evolved over all these years without a solid set of good management measurement practices. Perhaps, historically this may have been the casualty of rapid technology advancement, too much attention on technology for technology's sake or simply too great a divide between IT and the business.

This is something that needs to change now. Extremely poor IT decision making has been taking place, especially within the last several years that threatens the very foundations for how IT will ever successfully support the business. Many decisions in terms of off-shoring, elimination of experienced staff, failed IT initiatives and an increase in the number of IT solutions that cannot be effectively operated and supported day-to-day are being made out of pure management frustration with how IT operates. For senior business executives, it boils down to this:

> *"We have no idea how effectively IT operates and whether they even understand that – the only metric we have to go on is cost and we're taking care of that…"*

IT simply cannot continue to operate without learning how to effectively govern itself. If this continues, IT service quality will continue to erode at a rapid rate and IT labor costs will soar as increasing effort is needed to handle non-value labor related to poor IT support and delivery practices.

Here are the common excuses you hear a lot within IT organizations for NOT doing metrics:

**Excuse #1:**

*"We have other priorities..."*

What priorities? How do you know what they really are? If you're not measuring, how do you know where to prioritize? What makes IT different from every other part of the business that does measure itself? What you are really saying here is "...we know what is important for the business and will decide for them without need for any validating facts..." (This is generally not recommended for a healthy career path).

**Excuse #2:**

*"We're very uncomfortable with exposing the levels of our performance with others in the organization..."*

In other words, there is a desire to keep operating in a manner that IT is comfortable with even though it may be a frustration for everyone in the company outside of IT. Better yet, continue to keep executive management blind to what is going on in IT so they can continue to make poor decisions concerning IT.

**Excuse #3:**

*"We don't have the tools to accurately measure and will wait until management provides funds for this…"*

In other words, if we can't be perfect, let's not do it at all. Blame management for not providing enough resources. IT is saying: "We can't think of any indicators to measure in the short term and are totally dependent on a tool to tell us what to measure".

Here's an interesting trick that has been seen many times in many IT organizations:

1) Measure performance based on some small number of indicators that reflect (even if not entirely accurate) key ITSM performance behaviors

2) Issue these in a regular report to senior executives

3) Senior executives will get excited about the kind of results being shown

4) Senior executives will then ask how these could be more accurate or provide further detail

5) IT will tell them further support is needed in the way of a tool

6) Senior executives will fund the tool

Think this doesn't work? Watch what happens the next time a major tool vendor wanders into your organization to sell an IT Dashboard solution. For many executives, that is the first time they will see anything resembling management metrics in IT and they tend to get very excited.

The bottom line here is that without any ITSM measurements that matter, service improvement simply cannot take place. Remember the following guiding principles when encountering any resistance to building and using metrics:

*"If you don't measure it, you can't manage it"*

*"If you don't measure it, you can't improve it"*

*"If you don't measure it, you probably don't care'"*

*"If you can't influence it, then don't measure it"*

## *2.4 Benefits of Using ITSM Metrics That Matter*

There are a many good reasons for building and implementing an ITSM metrics program. Metrics that matter will provide:

- ✓ Senior executives and management with indicators from which they can make accurate and timely business decisions
- ✓ Visibility into how effectively and efficiently IT support and delivery services truly operate
- ✓ A basis for identifying and prioritizing IT service improvement enhancements
- ✓ Analytical information to identify service deficiencies and problems before they result in serious impacts
- ✓ A process-based focus for getting at root cause of deficiencies in service operations versus finger-pointing and blaming specific workers
- ✓ Senior management with confidence that IT is managing itself well

In addition, a set of Metrics That Matter can also be the foundation for modeling the impacts of business and IT decisions. For example, the impact of a business acquisition can be modeled in terms of an increase to specific operational metrics like the number of changes anticipated. This in turn can be calculated into the Change Efficiency Rate to see if it trips past an acceptable Tolerance level.

Lastly, the ultimate reason for instituting a Metrics That Matters program is to prevent operational risk. The following outcomes identify the kinds of things that IT should be trying to avoid as part of effective service support and delivery practices:

- ✓ Legal Exposure
- ✓ Service Outages
- ✓ Rework
- ✓ Waste
- ✓ Delayed Solutions
- ✓ Slow Operational Processes
- ✓ Security Breaches
- ✓ Inaccurate Information
- ✓ Slow Turnaround Times
- ✓ Unexpected Costs
- ✓ Higher or escalating costs
- ✓ Low Employee Morale
- ✓ Slow Response to Business Needs and Changes
- ✓ Unwanted PR Exposure
- ✓ Dissatisfied Customers
- ✓ Dissatisfied Suppliers
- ✓ Inability to scale
- ✓ Fines and Penalties
- ✓ High Levels of Non-Value Labor
- ✓ Loss of Market Share
- ✓ Loss of Revenue/Sales

An effective metrics measurement program needs to continually communicate the risk exposure levels to the events shown on the previous page to senior management based on the impacts of business and IT decisions. An approach for doing this is offered by use of the ITSM Metrics Modeling tool included with this book.

These kinds of communications are critical between IT and the senior leadership of the company. They go a long way towards fostering confidence and ability in IT to manage itself well and proactively take actions based on business needs and priorities.

# Chapter 3

# The ITSM Metrics Model

## *3.1 Categories of Metrics*

The ITSM Metrics Model uses several metric categories that are integrated into an overall metrics framework. These categories are as follows:

- Operational
- Key Performance Indicators (KPIs)
- Tolerances
- Critical Success Factors (CSFs)
- Dashboards
- Outcomes
- What-Ifs
- Analytical
- Other

These categories interact with each other in a manner that translates observations of operational events into indicators that can be used to make key IT and business management decisions. A model of these can be shown as follows:

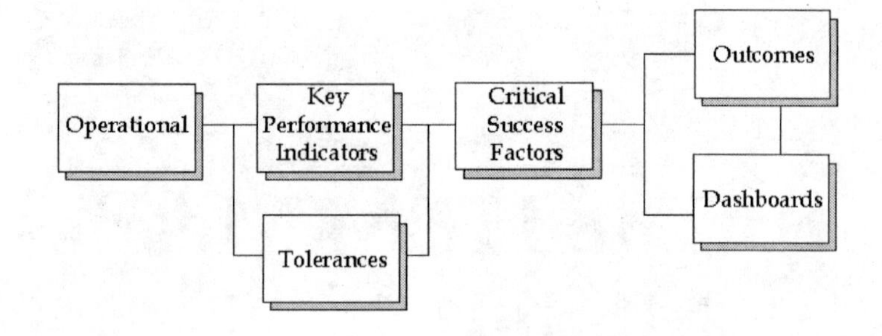


IT Service management Metrics Model

As shown by the above, *Operational* Metrics will be calculated into *KPIs (Key Performance Indicators)*. *KPI* results will fall into *Tolerance* ranges. *KPIs* are then calculated into *CSFs (Critical Success Factors)*. *CSFs* are then used to determine *Outcomes (Operational Risks)* and *Dashboard* measures.

Later chapters in this book provide a suggested list of ITSM metrics for each ITIL process for each of these metric categories. The modeling tool included with this book is actually based around this model as well.

The metrics suggested in this book are by no means meant to represent a complete list. They can be used as a starting point for which you may use them as a springboard for identifying additional metrics that your business or IT organization may find useful.

## *3.2 Operational Metrics*

These are basic observations of operational events for each ITSM process area. They are the starting point for the model and will be used to calculate the KPIs. Examples of these are as follows:

- Total Number Of Changes Implemented
- Number Of Incidents Reopened
- Number Of Problems In Pipeline
- Number Of Calls Handled
- Customer Satisfaction Rating
- Total Expended IT Costs

Inputs for these can come from a variety of places such as a Change Management System, Incident Management System, Service Desk ACD Reports, Surveys and other means.

## *3.3 Key Performance Indicators (KPIs)*

These are metrics that are used to indicate the performance level of an operation or process. KPIs are used to provide a basis for actionable management decisions. While Operational Metrics are generally historical in nature, KPIs are really the "Metrics That Matter".

KPIs are calculated or derived from one or more Operational Metrics. The results of these calculations are then compared to a Tolerance range to identify whether those results fall within acceptable levels. Examples of KPIs are as follows:

- Change Efficiency Rate
- Change Labor Workforce Utilization
- Incident Repeat Rate
- Capacity Management Process Maturity
- Total Service Penalties Paid

The above examples may not be clear understood purely by their names. Usually these require a small definition or explanation such that the KPI is understood. For this reason, KPIs and their associated calculations should be documented and agreed to by IT and Business Management.

In the above example, the *Change Efficiency Rate* is defined as "…a rate that indicates how many IT Changes put into the pipeline for the current month actually were implemented in the current month…" In other words, this KPI is meant to identify how efficient the organization is at processing changes. It would be calculated as:

*Total Changes Implemented / Total Changes in Pipeline*

With the above calculation, the dividend and divisor are taken from the Operational Metrics.

Not all KPIs require a calculation. The Capacity Management Process Maturity, for example, is an Operational Metric (observed from a process audit) and simply carries over as a KPI.

## *3.4 Tolerances*

Tolerances represent upper and lower boundaries for acceptable and non-acceptable KPI values. They should be set by the IT Service Manager and agreed to by IT and Business Senior Management. These are critical, as they form the basis for when management needs to take action or make a key decision.

Each KPI should be associated with one or more Tolerance values. For example, an upper value can represent a desired service target for the KPI and a lower value can represent a warning level or point at which some further action should occur.

The following table shows some examples of KPIs along with their associated Tolerance values:

| KPI | Service Target | Warning Level |
|---|---|---|
| Change Efficiency Rate | 92.5% | 85.0% |
| Customer Satisfaction Level | 8.7 | 7.9 |
| Average Incident Resolution Hours | 2.0 | 3.5 |
| Capacity Mgt Process Maturity | 3.0 | 2.5 |

In the above example, the service target for the *Change Efficiency Rate* would be 92.5%. Performance of that KPI would be considered acceptable as long as it did not fall lower than 85.0%. If it does fall lower, management actions may need to take place to raise the performance back to acceptable levels.

Note that Tolerance values are based on desired service and performance levels that the business is willing to tolerate. These can vary from one business versus another business.

## *3.5 Critical Success Factors (CSFs)*

These are metrics that represent key operational performance requirements which indicate whether a process or operation is performing successfully from a customer or business perspective. They are calculated or derived from one or more KPIs by comparing how those KPIs performed within the tolerance range.

A CSF is usually indicated with a performance level that is indicates a likelihood of success as to whether the CSF was achieved. Typically, this performance level can be something as simple as *High*, *Medium* or *Low*. Examples of these might look as follows:

| CSF | Performance Level |
|---|---|
| Protect Services When Making Changes | High |
| Provide Services At Acceptable Costs | Medium |
| Continually Improve Availability Of Services | Low |

A recommended approach for deriving a CSF is to first identify which KPIs relate to it and then rate the CSF based on the lowest value observed in any one of those KPIs. Using the above for example:

*Protect Services When Making Changes* might be derived from the Emergency Change Rate, Unauthorized Change Rate and Change Incident Rate KPIs. These all relate to how successful the organization is in meeting that CSF. To receive the performance level of *High*, all KPIs must have met or exceeded their Tolerance service targets.

However, let's say one of the KPIs falls below a service target Tolerance. In this case, the CSF performance level might be *Medium* or *Low* depending on how the KPI value fell within the specified Tolerance range for it.

## *3.6 Dashboards*

These are key metrics that are represented on a report or graphical interface that indicate the success, at risk or failure of a business operation. They are used to quickly assess the state of operation and take timely actions to correct operational deficiencies. In the ITSM Metrics Model presented here, Dashboard results are derived from CSF results.

Dashboards are generally used by management as a quick and easy way to spot deficiencies without wading through lots of reporting detail. They typically provide information at very high levels and may include drill down capabilities to see things in more detail.

Dashboards come in all forms, shapes and sizes. For the purposes of ITSM "Metrics That Matter", something called the Balanced Scorecard is suggested here. The Balanced Scorecard is an approach originally developed in the 1990's by Dr. Robert Kaplan and David Norton from the Harvard Business School. It was originally developed around the concept that financial measures alone are not critical for business success.

The Balanced Scorecard has been around for some time and is generally recognized as an acceptable approach for senior management levels. The scorecard categories recommended for ITSM are:

- Customer
- Capabilities
- Operational
- Financial
- Regulatory

*Customer*

The Customer category represents the customer view of the services being delivered. Are they satisfied? Are they serviced in accordance with agreements and expectations? Examples of some Change Management CSFs that contribute to Customer might be:

- Protect Services When Making Changes
- Make Changes Quickly And Accurately In Line With Business Needs

Both of these CSFs impact how a customer might be receiving (or not receiving) their services.

*Capabilities*

The Capabilities category represents, in the ITSM sense, the capability of the IT organization to meet business needs. Is there enough capacity to handle planned business volumes? Is there enough capacity to handle anticipated business and IT changes? Does the IT staff possess the right skills? Examples of some Capacity Management CSFs that contribute to Capabilities might be:

- Provide Services With Appropriate Capacity To Match Business Need
- Provide Accurate Capacity Forecasts

The above CSFs represent whether the IT organization is capable of delivering needed capacity to support services by accurately predicting capacity needs and providing needed capacity at the right time to match business requirements.

*Operational*

The Operational category represents, in the ITSM sense, how well the IT organization is delivering their services on a day-to-day basis. Are service levels being met? Are incidents resolved on a timely basis? Examples of some Incident Management CSFs that contribute to Operational might be:

- Quickly Resolve Incidents
- Maintain IT Service Quality

The above CSFs relate to every day tasks (in this case Incident Management tasks) and whether those tasks are operating in a repeatable, consistent, efficient and effective manner to quickly resolve incidents and take actions to maintain the quality of the services being delivered.

*Financial*

The Financial category represents, in the ITSM sense, how well the IT organization is managing and controlling costs as well as protecting and enhancing revenue. Are IT costs effectively managed? Are costs staying within budget? Does revenue received for IT chargeback cover the costs for the services being charged for? Examples of some Financial Management CSFs that contribute to Financial might be:

- Provide Effective Stewardship Of IT Finances
- Maintain Overall Effectiveness Of The IT Financial Management Process
- Recapture IT Costs Through Chargeback For Delivery Of It Services

*Regulatory*

The Regulatory category represents, in the ITSM sense, how well the IT organization is operating in a manner that protects it against regulatory risks for fines, penalties and audit issues. While not part of the original Balanced Scorecard approach, it has been included here because recent emphasis on IT regulatory issues within the last several years.

Sample questions from this category might include: Is effective stewardship maintained over IT costs? Is the infrastructure protected from unauthorized changes? Is the infrastructure adequately protected from security risks? Examples of some CSFs that contribute to Regulatory might be:

- Provide Effective Stewardship Of IT Finances
- Utilize A Repeatable Process For Handling Changes
- Provide A Repeatable Process For Rolling Out Releases
- Maintain viability of IT Service Continuity Plans

As can be seen from the examples provided, CSFs can contribute to one or more scorecard areas. Likewise, each scorecard area may have one or multiple CSFs.

## *3.7 Outcomes*

Outcomes are key indicators of general business risk areas. These are associated with performance indicators that identify the success, at risk or failure of KPIs or CSFs. Outcomes are used to quickly assess the level of risks created by process or operational deficiencies. In short, outcomes are the kind of things that IT is trying to protect against. Examples of these include:

- ✓ Legal Exposure
- ✓ Service Outages
- ✓ Rework
- ✓ Waste
- ✓ Security Breaches
- ✓ Unexpected Costs
- ✓ Slow Response To Business Needs And Changes
- ✓ Unwanted PR Exposure
- ✓ Dissatisfied Customers
- ✓ Fines and Penalties
- ✓ High Levels Of Non-Value Labor
- ✓ Loss of Market Share
- ✓ Loss of Revenue/Sales

Each of the above Outcomes can be associated with a performance indicator such as *High, Medium* or *Low* that might reflect the likelihood of risk that the Outcome will occur. In the ITSM model, the risk level is derived from the mean average of the CSF performance levels.

As an example, the following CSFs relate to risks of incurring *Service Outages*:

- ✓ Quickly Resolve Incidents
- ✓ Minimize The Impact Of Problems (Reduce Incident Frequency/Duration)
- ✓ Protect Services When Making Changes
- ✓ Implement High Quality Releases
- ✓ Protect Services From Capacity Related Incidents

Scoring for an Outcome runs opposite to how the CSFs are calculated. If a CSF scores *Low*, meaning the likelihood of achieving that CSF is low, then the Outcome would score *High*. This means that the risk of the Outcome occurring is high because the CSF achievement was low.

### 3.8 *What-Ifs*

What-Ifs can be characterized as Use Cases derived from impending business decisions. These will be used to "model" the impacts of those decisions on KPIs and CSFs. An example of a Use Case is simply a scenario for some future event. Examples might include:

- What happens if a major new application goes into production?
- What happens if a planned merger or acquisition occurs?
- What happens if we cut operational staff?

Each of the above is examples of a future event or business decision that IT or business executive management might be thinking of.

For each of these, you can model the impact of such an event by raising or decreasing the values of the Operational Indicators that might be related to them. For example: let's say a new application is going into production. You may decide to model this by:

- ✓ Increasing *the Number of Releases in the Pipeline*
- ✓ Increasing the *Number of Labor Hours Spent On Releases*
- ✓ Increasing the *Number of Changes in the Pipeline*
- ✓ Increasing the *Number of Incidents* by some factor such as 30%

Once these changes are made, you can then examine the impact on KPI and CSF results to see if they fall out of Tolerance levels. These kinds of changes reflect your best guess as to how a business decision may impact the quality of your service support and delivery capabilities. Quite a powerful technique to use with executive management!

## *3.9 Analytical*

The Analytical category of metrics is used to separate out certain metrics that are really more helpful for supporting research into an issue, incident or service problem. These are metrics that you may report on only on a one-time basis or as part of a drill-down (such as for a Dashboard).

Typically, these kinds of metrics are usually subsets or subdivisions of other metrics. One example might be the Operational Metric of *Total Number of Incidents.* For analytical purposes, you may also wish to see this total broken out by:

- Geographic Region
- Department or Business Unit
- Technology Platform
- IT Service Delivered
- Time of Day, etc.

Recognize that the more of these you include the more complex and labor intensive your ITSM Metrics Program will become. IT frequently makes the mistake of including these in regular reporting to senior management "just in case". This results in a lot of wasted labor in building reports and clouds real management issues that need to be addressed.

Try to avoid as much of this as possible. If pressed to include these, try to keep them in separate reports or as part of a drill-down presentation whenever possible. These are important for when you are looking where to make improvements, but recognize they are important *only* when in the process of making those improvements.

## *3.10 Other*

This category is reserved for other kinds of metrics that don't quite fall into the earlier categories. Examples of these might include:

- ✓ Indicators or other observed events that are being used to represent other metrics that can't be accurately reported on (see Alternatives If Few Metrics chapter for these).
- ✓ Resource and Service metrics such as server utilizations or transaction response times.
- ✓ Business metrics such as headcounts, customer and sales volumes or planned revenues

The following chapters in this book contain suggested metrics for each ITIL process plus Service Desk and Work Force Management functions. They also show an inventory of recommended Operational, KPI, CSF and associated calculations for each one. These are not meant to be exhaustive, but it is hoped they will kindle ideas within your own organization as to what should be part of your ITSM metrics program.

While not an exhaustive list of metrics by any means, it is strongly felt that if only the metrics listed in this book were tracked and modeled, simple as they are, IT will still be greatly much better off than where many organizations stand today.

# Chapter 4

# Incident Management Metrics

## *4.1 Operational Metrics*

The following table lists suggested Incident Management Operational Metrics.

| XREF | METRIC |
|---|---|
| A | Total Number Of Incidents |
| B | Average Time To Resolve Severity 1 and Severity 2 Incidents (Hours) |
| C | Number Of Incidents Resolved Within Agreed Service Levels |
| D | Number Of High Severity/Major Incidents |
| E | Number Of Incidents With Customer Impact |
| F | Number Of Incidents Reopened |
| G | Total Available Labor Hours To Work On Incidents (Non-Service Desk) |
| H | Total Labor Hours Spent Resolving Incidents (Non-Service Desk) |
| I | Incident Management Tooling Support Level |
| J | Incident Management Process Maturity |

Suggested sources for Incident Management Operational Metrics can be found in places such as:

- ✓ Incident Management System Reports
- ✓ Labor or Other HR Reports
- ✓ Process and Tool Assessment Audit Findings

## *4.2 Key Performance Indicators (KPIs)*

The following table lists suggested KPIs and how they are calculated from the Operational Metrics listed previously.

| XREF | KPI | CALCULATION |
|---|---|---|
| 1 | Number Of Incident Occurrences | *A* |
| 2 | Number Of High Severity/Major Incidents | *D* |
| 3 | Incident Resolution Rate | *C/A* |
| 4 | Customer Incident Impact Rate | *E/A* |
| 5 | Incident Reopen Rate | *F/A* |
| 6 | Average Time To Resolve Severity 1 and Severity 2 Incidents (Hours) | *B* |
| 7 | Incident Labor Utilization Rate | *H/G* |
| 8 | Incident Management Tooling Support Level | *I* |
| 9 | Incident Management Process Maturity | *J* |

## *4.3 Why These Metrics (KPIs) Matter*

The KPIs described earlier are critical to managing and controlling Incident Management activities. The following table lists each Incident Management KPI and the question it is trying to answer:

| KPI | Question Being Answered |
|---|---|
| Number Of Incident Occurrences | How many incidents did we experience within our infrastructure? |
| Number Of High Severity/Major Incidents | How many major incidents did we experience? |
| Incident Resolution Rate | How successful are we at resolving incidents per business requirements? |
| Customer Incident Impact Rate | How well are we keeping incidents from impacting customers? |
| Incident Reopen Rate | How successful are we at permanently resolving incidents? |
| Average Time To Resolve Severity 1 and Severity 2 Incidents (Hours) | How quickly are we resolving incidents? |
| Incident Labor Utilization Rate | How much available labor capacity was spent handling incidents? |
| Incident Management Tooling Support Level | How well does our current tool set support Incident Management activities? |
| Incident Management Process Maturity | How well do we execute our Incident Management practices? |

## *4.4 Critical Success Factors (CSFs)*

The table below lists suggested Critical Success Factors for Incident Management. The KPI references listed in the right column indicate which KPIs are used as input for the associated CSF.

| CSF | KPI |
|---|---|
| Quickly Resolve Incidents | *5,6,8* |
| Maintain IT Service Quality | *1,2,3,4,8,9* |
| Improve IT And Business Productivity | *7,8* |
| Maintain User Satisfaction | *4,8,9* |

# Chapter 5

# Problem Management Metrics

## *5.1 Operational Metrics*

The following table lists suggested Problem Management Operational Metrics.

| XREF | METRIC |
|---|---|
| A | Number Of Repeat Incidents |
| B | Number Of Major Problems |
| C | Total Number Of Incidents |
| D | Total Number Of Problems In Pipeline |
| E | Number Of Problems Removed (Error Control) |
| F | Number Of Known Errors (Root Cause Known and Workaround In Place) |
| G | Number Of Problems Reopened |
| H | Number Of Problems With Customer Impact |
| I | Average Problem Resolution Time - Severity 1 and 2 Problems (Days) |
| J | Total Available Labor Hours To Work On Problems |

| XREF | METRIC |
|---|---|
| K | Total Labor Hours Spent Working On And Coordinating Problems |
| L | Problem Management Tooling Support Level |
| M | Problem Management Process Maturity |

Suggested sources for Problem Management Operational Metrics can be found in places such as:

- ✓ Incident Management System Reports
- ✓ Problem Management System Reports
- ✓ Labor or Other HR Reports
- ✓ Process and Tool Assessment Audit Findings

## *5.2 Key Performance Indicators (KPIs)*

The following table lists suggested KPIs and how they are calculated from the Operational Metrics listed previously.

| XREF | KPI | CALCULATION |
|---|---|---|
| 1 | Incident Repeat Rate | *A/C* |
| 2 | Number Of Major Problems | *B* |
| 3 | Problem Resolution Rate | *E/D* |
| 4 | Problem Workaround Rate | *F/A* |
| 5 | Problem Reopen Rate | *F/D* |
| 6 | Customer Impact Rate | *H/D* |
| 7 | Average Problem Resolution Time - Severity 1 and 2 Problems (Days) | *I* |
| 8 | Problem Labor Utilization Rate | *K/J* |
| 9 | Problem Management Tooling Support Level | *L* |
| 10 | Problem Management Process Maturity | *M* |

## *5.3 Why These Metrics (KPIs) Matter*

The KPIs described earlier are critical to managing and controlling Problem Management activities. The following table lists each Problem Management KPI and the question it is trying to answer:

| KPI | Question Being Answered |
|---|---|
| Incident Repeat Rate | How effective are we at minimizing repeat incidents? |
| Number Of Major Problems | How many major problems did we experience? |
| Problem Resolution Rate | What percentage of problems have we eliminated? |
| Problem Workaround Rate | For what percentage of problems did we implement workarounds? |
| Problem Reopen Rate | How successful are we at removing problems permanently? |
| Customer Impact Rate | How well are we keeping problems from impacting customers? |
| Average Problem Resolution Time - Severity 1 and 2 Problems (Days) | How long does it take us to resolve problems? |
| Problem Labor Utilization Rate | How much available labor capacity was spent handling problems? |
| Problem Management Tooling Support Level | How well does our current tool set support Problem Management activities? |

| KPI | Question Being Answered |
|---|---|
| Problem Management Process Maturity | How well do we execute our Problem Management practices? |

## *5.4 Critical Success Factors (CSFs)*

The table below lists suggested Critical Success Factors for Problem Management. The KPI references listed in the right column indicate which KPIs are used as input for the associated CSF.

| CSF | KPI |
|---|---|
| Minimize The Impact Of Problems (Reduce Incident Frequency/Duration) | *1,2,6,7,9* |
| Improve Quality Of Services Being Delivered | *1,2,10* |
| Resolve Problems and Errors Efficiently and Effectively | *3,4,5,8,9,10* |

# Chapter 6

# Change Management Metrics

## *6.1 Operational Metrics*

The following table lists suggested Change Management Operational Metrics.

| XREF | METRIC |
|---|---|
| A | Total Changes In Pipeline |
| B | Total Changes Implemented |
| C | Number Of Failed Changes |
| D | Number of Emergency Changes |
| E | Number of Unauthorized Changes Detected |
| F | Number of Changes Rescheduled |
| G | Average Process Time Per Change (Days) |
| H | Number of Changes Resulting In Incidents |
| I | Change Management Tooling Support Level |
| J | Change Management Process Maturity |
| K | Total Available Labor Hours To Coordinate (Not Implement) Changes |
| L | Total Labor Hours Spent Coordinating Changes |

Suggested sources for Change Management Operational Metrics can be found in places such as:

- ✓ Change Management System Reports
- ✓ Incident Management System Reports
- ✓ Labor or Other HR Reports
- ✓ Process and Tool Assessment Audit Findings
- ✓ Observations of Incidents or CMDB/Asset Reports to Detect Unauthorized Changes

## *6.2 Key Performance Indicators (KPIs)*

The following table lists suggested KPIs and how they are calculated from the Operational Metrics listed previously.

| XREF | KPI | CALCULATION |
|---|---|---|
| 1 | Change Efficiency Rate | *B/A* |
| 2 | Change Success Rate | *1-(C/B)* |
| 3 | Emergency Change Rate | *D/A* |
| 4 | Change Reschedule Rate | *F/A* |
| 5 | Average Process Time Per Change (Days) | *G* |
| 6 | Unauthorized Change Rate | *E/B* |
| 7 | Change Incident Rate | *H/B* |
| 8 | Change Labor Workforce Utilization | *L/K* |
| 9 | Change Management Tooling Support Level | *I* |
| 10 | Change Management Process Maturity | *J* |

## *6.3 Why These Metrics (KPIs) Matter*

The KPIs described earlier are critical to managing and controlling Change Management activities. The following table lists each Change Management KPI and the question it is trying to answer:

| KPI | Question Being Answered |
|---|---|
| Change Efficiency Rate | How efficient are we at handling changes? |
| Change Success Rate | How effective are we at handling changes? |
| Emergency Change Rate | What percentage of changes were emergencies? |
| Change Reschedule Rate | How well do we implement changes on schedule? |
| Average Process Time Per Change (Days) | How long does the average change take? |
| Unauthorized Change Rate | What percentage of changes bypassed the Change process? |
| Change Incident Rate | What percentage of changes caused incidents? |
| Change Labor Workforce Utilization | How much available labor capacity was used to handle and coordinate changes? |
| Change Management Tooling Support Level | How well does our current tool set support Change Management activities? |
| Change Management Process Maturity | How well do we execute our Change Management practices? |

## *6.4 Critical Success Factors (CSFs)*

The table below lists suggested Critical Success Factors for Change Management. The KPI references listed in the right column indicate which KPIs are used as input for the associated CSF.

| CSF | KPI |
|---|---|
| Protect Services When Making Changes | *3,6,7* |
| Make Changes Quickly And Accurately In Line With Business Needs | *4,5,7,8,9* |
| Make Changes Efficiently And Effectively | *1,2,5,9* |
| Utilize A Repeatable Process For Handling Changes | *3,6,9,10* |

# Chapter 7

# Release Management Metrics

## 7.1 Operational Metrics

The following table lists suggested Release Management Operational Metrics.

| XREF | METRIC |
|---|---|
| A | Total Releases In Pipeline |
| B | Total Releases Implemented |
| C | Number Of Failed Releases |
| D | Number of Releases Rescheduled |
| E | Average Process Time Per Release |
| F | Number of Releases Resulting In Incidents |
| G | Release Management Tooling Support Level |
| H | Release Management Process Maturity |
| I | Total Available Labor Hours To Implement Releases |
| J | Total Labor Hours Spent Implementing Releases |
| K | Number of Known Release Errors In Production |
| L | Number of Releases Withdrawn |

Suggested sources for Release Management Operational Metrics can be found in places such as:

- ✓ Release Management System Reports
- ✓ Incident Management System Reports
- ✓ Labor or Other HR Reports
- ✓ Process and Tool Assessment Audit Findings
- ✓ Project Management Reports
- ✓ Project Status Reports

## *7.2 Key Performance Indicators (KPIs)*

The following table lists suggested KPIs and how they are calculated from the Operational Metrics listed previously.

| XREF | KPI | CALCULATION |
|---|---|---|
| 1 | Release Efficiency Rate | *B/A* |
| 2 | Release Success Rate | *1-(C/B)* |
| 3 | Release Reschedule Rate | *D/A* |
| 4 | Release Defect Rate | *F/B* |
| 5 | Release Labor Utilization | *J/I* |
| 6 | Release Management Tooling Support Level | *G* |
| 7 | Release Management Process Maturity Level | *H* |
| 8 | Number of Known Release Errors In Production | *K* |
| 9 | Release Withdraw Rate | *L/A* |
| 10 | Release Labor Waste Rate | *(C+F+L)/A* |

## *7.3 Why These Metrics (KPIs) Matter*

The KPIs described earlier are critical to managing and controlling Release Management activities. The following table lists each Release Management KPI and the question it is trying to answer:

| KPI | Question Being Answered |
|---|---|
| Release Efficiency Rate | How efficient are we at handling releases? |
| Release Success Rate | How successful are we at implementing releases? |
| Release Reschedule Rate | How well do we implement releases on schedule? |
| Release Defect Rate | What percentage of releases caused incidents? |
| Release Labor Utilization | How much labor capacity was used to handling releases? |
| Release Management Tooling Support Level | How well does our current tool set support Release Management activities? |
| Release Management Process Maturity Level | How well do we execute our Release Management practices? |
| Number of Known Release Errors In Production | How good is the quality of releases put into production? |
| Release Withdraw Rate | What percent of releases never go into production? |
| Release Labor Waste Rate | What percent of release labor is wasted? |

## *7.4 Critical Success Factors (CSFs)*

The table below lists suggested Critical Success Factors for Release Management. The KPI references listed in the right column indicate which KPIs are used as input for the associated CSF.

| CSF | KPI |
|---|---|
| Provide Repeatable Process For Rolling Out Releases | *6,7* |
| Implement High Quality Releases | *2,4,6,8* |
| Implement Releases Efficiently And Effectively | *1,2,3,5,6,9,10* |

# Chapter 8

# Configuration Management Metrics

## *8.1 Operational Metrics*

The following table lists suggested Configuration Management Operational Metrics.

| XREF | METRIC |
|---|---|
| A | Total Number of CIs in CMDB |
| B | Number of CIs Audited |
| C | Number of CI Errors Discovered |
| D | Configuration Management Tooling Support Level |
| E | Configuration Management Process Maturity |
| F | Number of CI Changes |
| G | Number of CI Changes Without Corresponding RFC |
| H | Number of Incidents Related To Inaccurate CI Information |

| XREF | METRIC |
|---|---|
| I | Number of Change Failures Related To Inaccurate CI Information |
| J | Number of Services Operating With Incomplete CI Information |
| K | Number Of Services In Service Catalog |
| L | Number of CIs Without Assigned Ownership |

Suggested sources for Configuration Management Operational Metrics can be found in places such as:

- ✓ CMDB Audit and Status Accounting Reports
- ✓ Incident Management System Reports
- ✓ Change Management System Reports
- ✓ Process and Tool Assessment Audit Findings
- ✓ Service Catalog Listings
- ✓ Auto-Discovery Reports

## *8.2 Key Performance Indicators (KPIs)*

The following table lists suggested KPIs and how they are calculated from the Operational Metrics listed previously.

| XREF | KPI | CALCULATION |
|---|---|---|
| 1 | CMDB Accuracy Ratio | *1-(C/A)* |
| 2 | Number of Incidents Related To Inaccurate CI Information | *H* |
| 3 | Number of Change Failures Related To Inaccurate CI Information | *I* |
| 4 | Configuration Management Tooling Support Level | *D* |
| 5 | Configuration Management Process Maturity | *E* |
| 6 | CMDB Completeness Ratio | *1-(J/K)* |
| 7 | CI Ownership Rate | *1-(L/A)* |

## *8.3 Why These Metrics (KPIs) Matter*

The KPIs described earlier are critical to managing and controlling Configuration Management activities. The following table lists each Configuration Management KPI and the question it is trying to answer:

| KPI | Question Being Answered |
|---|---|
| CMDB Accuracy Ratio | How accurate is information in the CMDB? |
| Number of Incidents Related To Inaccurate CI Information | How many incidents were related to inaccurate configuration information? |
| Number of Change Failures Related To Inaccurate CI Information | How many changes failed due to inaccurate configuration information? |
| Configuration Management Tooling Support Level | How well does our current tool set support Configuration Management activities? |
| Configuration Management Process Maturity | How well do we execute our Configuration Management practices? |
| CMDB Completeness Ratio | How complete is our configuration information? |
| CI Ownership Rate | How much of our infrastructure has no assigned ownership? |

## *8.4 Critical Success Factors (CSFs)*

The table below lists suggested Critical Success Factors for Configuration Management. The KPI references listed in the right column indicate which KPIs are used as input for the associated CSF.

| CSF | KPI |
|---|---|
| Control Information About The IT Infrastructure | *1,4,5* |
| Support Delivery Of Quality IT Services | *2,3,4,6,7* |

# Chapter 9

# Service Desk Metrics

## *9.1 Operational Metrics*

The following table lists suggested Service Desk Operational Metrics.

| XREF | METRIC |
|---|---|
| A | Total Number Of Calls To Service Desk |
| B | Average Call Duration (Minutes) |
| C | Average Call Waiting (Minutes) |
| D | Service Desk Tooling Support Level |
| E | Number of Calls Transferred |
| F | Number of Calls Abandoned |
| G | Available Call Agent Labor Hours |
| H | Total Service Desk Available Hours |
| I | Total Service Desk Unavailable Hours |

Suggested sources for Service Desk Operational Metrics can be found in places such as:

- ✓ Automatic Call Distribution (ACD) Reports
- ✓ Incident Management System Reports
- ✓ Service Level Agreements (SLAs)
- ✓ Tool Assessment Audit Findings

## *9.2 Key Performance Indicators (KPIs)*

The following table lists suggested KPIs and how they are calculated from the Operational Metrics listed previously.

| XREF | KPI | CALCULATION |
|---|---|---|
| 1 | Service Desk Call Resolution Rate | *1-((E+F)/A)* |
| 2 | Average Call Duration (Minutes) | *B* |
| 3 | Service Desk Tooling Support Level | *D* |
| 4 | Call Agent Utilization | *((B*A)/60)/G* |
| 5 | Call Abandon Rate | *F/A* |
| 6 | Call Duration Experience | *B+C* |
| 7 | Call Waiting Rate | *C/(C+B)* |
| 8 | Service Desk Service Availability | *1-(I/J)* |

## *9.3 Why These Metrics (KPIs) Matter*

The KPIs described earlier are critical to managing and controlling Service Desk activities. The following table lists each Service Desk KPI and the question it is trying to answer:

| KPI | Question Being Answered |
|---|---|
| Service Desk Call Resolution Rate | How many calls are resolved at the Service Desk? |
| Average Call Duration (Minutes) | How long is the average customer call experience? |
| Service Desk Tooling Support Level | How well does our current tool set support Service Desk activities? |
| Call Agent Utilization | Do we have enough resources to handle calls? |
| Call Abandon Rate | What percent of callers hang up before getting service? |
| Call Duration Experience | What is the total time taken to service callers? |
| Call Waiting Rate | What percentage of total caller time is spent waiting? |
| Service Desk Service Availability | Is the Service Desk available when needed? |

## *9.4 Critical Success Factors (CSFs)*

The table below lists suggested Critical Success Factors for the Service Desk function. The KPI references listed in the right column indicate which KPIs are used as input for the associated CSF.

| CSF | KPI |
|---|---|
| Resolve Customer Issues And Problems At First Call | *1,3* |
| Maintain Customer Productivity | *1,3,6* |
| Provide A Positive Customer Call Experience | *3,5,6,7,8* |
| Provide Effective Support For Customer Calls | *3,4* |

# Chapter 10

# Service Level Management Metrics

## *10.1 Operational Metrics*

The following table lists suggested Service Level Management Operational Metrics.

| XREF | METRIC |
|---|---|
| A | Overall Customer Satisfaction Rating |
| B | Number Of Services Delivered To Customers/Business (SLAs) |
| C | Number Of Services Without SLAs |
| D | Number Of Internal Services Supporting SLAs (OLAs) |
| E | Number Of Internal Supporting Services Without OLAs |
| F | Number Of Supporting Services Delivered By Vendors |
| G | Number Of Vendor Delivered Services Without Agreed Service Targets |

| XREF | METRIC |
|---|---|
| H | Total Service Penalties Paid |
| I | Total Number Of SLA Service Targets |
| J | Total Number Of SLA Service Targets Breached |
| K | Number Of SLAs Operating Without Service Owners |
| L | Service Level Management Tooling Support Level |
| M | Service Level Management Process Maturity |

Suggested sources for Service Level Management Operational Metrics can be found in places such as:

- ✓ Customer Survey Results
- ✓ Service Catalog Listings
- ✓ Procurement Contract Files
- ✓ Accounts Payable Reports
- ✓ Service Level Agreements
- ✓ Service Level Reporting
- ✓ Service Level Management Database Reports
- ✓ Tool And Process Assessment Audit Findings

## *10.2 Key Performance Indicators (KPIs)*

The following table lists suggested KPIs and how they are calculated from the Operational Metrics listed previously.

| XREF | KPI | CALCULATION |
|---|---|---|
| 1 | Overall Customer Satisfaction Rating | *A* |
| 2 | SLA Coverage Rate | *1-(C/B)* |
| 3 | OLA Coverage Rate | *1-(E/D)* |
| 4 | Percent Of Vendor Services Delivered Without Agreed Service Targets | *G/F* |
| 5 | Total Service Penalties Paid | *H* |
| 6 | Percent Of SLA Service Targets Adhered To | *1-(J/I)* |
| 7 | Percent Of SLAs With Responsible Service Owners | *1-(K/B)* |
| 9 | Service Level Management Tooling Support Level | *L* |
| 10 | Service Level Management Process Maturity | *M* |

## *10.3 Why These Metrics (KPIs) Matter*

The KPIs described earlier are critical to managing and controlling Service Level Management activities. The following table lists each Service Level Management KPI and the question it is trying to answer:

| KPI | Question Being Answered |
|---|---|
| Overall Customer Satisfaction Rating | How do customers perceive the quality of the services we are delivering? |
| SLA Coverage Rate | What percent of the services we deliver have formally been agreed to? |
| OLA Coverage Rate | What percent of our internal supporting services operate with formal agreements? |
| Percent Of Vendor Services Delivered Without Agreed Service Targets | What percent of our supporting services are delivered without agreed service targets? |
| Total Service Penalties Paid | How much are we paying in service penalties? |
| Percent Of SLA Service Targets Adhered To | How well have we met our SLA service targets? |
| Percent Of SLAs With Responsible Service Owners | What percent of our delivered services do not have assigned service owners? |
| Service Level Management Tooling Support Level | How well does our current tool set support Service Level Management activities? |

| KPI | Question Being Answered |
|---|---|
| Service Level Management Process Maturity | How well do we execute our Service Level Management Processes? |

## *10.4 Critical Success Factors (CSFs)*

The table below lists suggested Critical Success Factors for Service Level Management. The KPI references listed in the right column indicate which KPIs are used as input for the associated CSF.

| CSF | KPI |
|---|---|
| Deliver IT Services As Agreed To By Customers And The Business | *2,3,4,5,6* |
| Manage The Business/User Interface | *1,7,8,9* |
| Provide Services At Acceptable Cost | *5* |
| Manage Quality Of IT Services In Line With Business Requirements | *1,5,6,8,9* |

# Chapter 11

# Availability Management Metrics

## *11.1 Operational Metrics*

The following table lists suggested Availability Management Operational Metrics.

| XREF | METRIC |
|---|---|
| A | Total Unplanned Expenses Related To Availability |
| B | Total Number of Incidents |
| C | Total Number Of Customer Impacting Incidents |
| D | Total Available Minutes For All Services Delivered |
| E | Total Unavailable Minutes For All Services Delivered |
| F | Availability Management Tooling Support Level |
| G | Availability Management Process Maturity Level |
| H | Total Number of Service Targets From Internal Suppliers |

| XREF | METRIC |
|---|---|
| I | Total Number of Service Targets From Vendor Suppliers |
| J | Number of Internal Supplier Targets Missed |
| K | Number of Vendor Supplier Targets Missed |
| L | Number of Security Related Incidents |
| M | Number of HW/SW/Networking CIs |
| N | Number of HW/SW/Networking CIs Not Supported By Vendors |
| O | Number Of Services In Service Catalog |
| P | Number of Services Not Covered By An Active Availability Plan |
| Q | Number of Services Without Availability Review Last 3 Months |

Suggested sources for Availability Management Operational Metrics can be found in places such as:

- ✓ Incident Management System
- ✓ Problem Management System
- ✓ Service Catalog Listings
- ✓ Service Level Agreements
- ✓ Operational Level Agreements
- ✓ Underpinning Contracts
- ✓ Configuration Management Database Reports
- ✓ Existing Availability Plans
- ✓ Service Level Reporting
- ✓ Service Level Management Database Reports
- ✓ Tool And Process Assessment Audit Findings

## *11.2 Key Performance Indicators (KPIs)*

The following table lists suggested KPIs and how they are calculated from the Operational Metrics listed previously.

| XREF | KPI | CALCULATION |
|---|---|---|
| 1 | Total Unplanned Expenses Related To Availability | *A* |
| 2 | Availability Resilience Index | *1-(C/B)* |
| 3 | Average Service Reliability Index | *1-(E/D)* |
| 4 | Availability Management Tooling Support Level | *F* |
| 5 | Availability Management Process Maturity Level | *G* |
| 6 | Average Internal Supplier Service Reliability Index | *1-(J/H)* |
| 7 | Average Vendor Supplier Service Reliability Index | *1-(K/I)* |
| 8 | Security Vulnerability Index | *L/B* |
| 9 | Serviceability Index | *N/M* |
| 10 | Availability Risk Index | *P/O* |
| 11 | Continuous Availability Improvement Index | *1-(Q/O)* |

## *11.3 Why These Metrics (KPIs) Matter*

The KPIs described earlier are critical to managing and controlling Availability Management activities. The following table lists each Availability Management KPI and the question it is trying to answer:

| KPI | Question Being Answered |
|---|---|
| Total Unplanned Expenses Related To Availability | How much unplanned costs were spent on maintaining needed availability? |
| Availability Resilience Index | How resilient is our infrastructure towards protecting services? |
| Average Service Reliability Index | How reliable are the services we deliver? |
| Availability Management Tooling Support Level | How well does our current tool set support Availability Management activities? |
| Availability Management Process Maturity Level | How well do we execute our Availability Management practices? |
| Average Internal Supplier Service Reliability Index | How reliably are internal suppliers supporting our services? |
| Average Vendor Supplier Service Reliability Index | How reliably are vendors suppliers supporting our services? |
| Security Vulnerability Index | How vulnerable are we to security threats? |
| Serviceability Index | How much of our physical infrastructure is supported by vendors? |

| KPI | Question Being Answered |
|---|---|
| Availability Risk Index | What percent of our services are delivered without addressing availability? |
| Continuous Availability Improvement Index | How well do we proactively look at improving service availability? |

## *11.4 Critical Success Factors (CSFs)*

The table below lists suggested Critical Success Factors for Availability Management. The KPI references listed in the right column indicate which KPIs are used as input for the associated CSF.

| CSF | KPI |
|---|---|
| Provide Services With Appropriate Availability To Match Business Need | *2,3,5,6,7,8,9,10* |
| Demonstrate Cost-Effectiveness Through Effective Availability Planning | *1,4* |
| Continually Improve Availability Of Delivered Services | *11* |

# Chapter 12

# Capacity Management Metrics

## 12.1 Operational Metrics

The following table lists suggested Capacity Management Operational Metrics.

| XREF | METRIC |
|---|---|
| A | Total Expenses For Unplanned Capacity |
| B | Number of IT Resource Forecasts |
| C | Number of IT Service Forecasts |
| D | Number of IT Business Forecasts |
| E | Number of Missed IT Resource Forecasts |
| F | Number of Missed IT Service Forecasts |
| G | Number of Missed IT Business Forecasts |
| H | Number of Incidents Caused By Inadequate Capacity |
| I | Total Actual IT Costs For Hardware, Software and Network |
| J | Capacity Management Tooling Support Level |
| K | Capacity Management Process Maturity Level |

Suggested sources for Capacity Management Operational Metrics can be found in places such as:

- ✓ Capacity Plans and Reports
- ✓ IT Financial Accounting and Budget Reports
- ✓ Incident Management System Reports
- ✓ Procurement Reports
- ✓ Tool And Process Assessment Audit Findings

## *12.2 Key Performance Indicators (KPIs)*

The following table lists suggested KPIs and how they are calculated from the Operational Metrics listed previously.

| XREF | KPI | CALCULATION |
|---|---|---|
| 1 | Total Expenses For Unplanned Capacity | *A* |
| 2 | Percent of IT Costs For Unplanned Capacity Expenses | *A/I* |
| 3 | IT Resource Forecast Accuracy Ratio | *1-(E/B)* |
| 4 | IT Service Forecast Accuracy Ratio | *1-(F/C)* |
| 5 | IT Business Forecast Accuracy Ratio | *1-(G/D)* |
| 6 | Number of Incidents Caused By Inadequate Capacity | *H* |
| 7 | Capacity Management Tooling Support Level | *J* |
| 8 | Capacity Management Process Maturity Level | *K* |

## *12.3 Why These Metrics (KPIs) Matter*

The KPIs described earlier are critical to managing and controlling Capacity Management activities. The following table lists each Capacity Management KPI and the question it is trying to answer:

| KPI | Question Being Answered |
|---|---|
| Total Expenses For Unplanned Capacity | How much did unplanned capacity cost us for HW/SW/Network Components? |
| Percent of IT Costs For Unplanned Capacity Expenses | What percent of our actual HW/SW/Network costs were for unplanned capacity? |
| IT Resource Forecast Accuracy Ratio | How accurate are we in forecasting IT Needed Resources? |
| IT Service Forecast Accuracy Ratio | How accurate are we in predicting workload volumes for services? |
| IT Business Forecast Accuracy Ratio | How accurate are we in anticipating business growth and changes? |
| Number of Incidents Caused By Inadequate Capacity | How many incidents were caused related to capacity? |
| Capacity Management Tooling Support Level | How well does our current tool set support Capacity Management activities? |
| Capacity Management Process Maturity Level | How well do we execute our Capacity Management practices? |

## *12.4 Critical Success Factors (CSFs)*

The table below lists suggested Critical Success Factors for Capacity Management. The KPI references listed in the right column indicate which KPIs are used as input for the associated CSF.

| CSF | KPI |
|---|---|
| Provide Accurate Capacity Forecasts | *3,4,5* |
| Provide Services With Appropriate Capacity To Match Business Need | *6,8* |
| Protect Services From Capacity Related Incidents | *6* |
| Demonstrate Cost-Effectiveness Through Accurate Capacity Planning | *1,2,7* |

# Chapter 13

# IT Service Continuity Management Metrics

## *13.1 Operational Metrics*

The following table lists suggested IT Service Continuity Management Operational Metrics.

| XREF | METRIC |
|---|---|
| A | Number Of Services In Service Catalog |
| B | Number of Services Covered By IT Service Continuity Plans |
| C | Number of Service Continuity Plan Audit Failures |
| D | Mean Time (Months) Between Continuity Tests For Each Service |
| E | Total Planned Labor Hours For IT Service Continuity Activities |
| F | Total Used Labor Hours For IT Service Continuity Activities |
| G | Number of IT Services Tested For Service Continuity |

| XREF | METRIC |
|---|---|
| H | Mean Time (Months) Between Continuity Plan Audits For Each Service |
| I | Total Planned Costs For ITSCM Activities |
| J | Total Unplanned Costs For ITSCM Activities |
| K | Number of Business Continuity Issues Logged Against ITSCM Plans |
| L | IT Service Continuity Tooling Support Level |
| M | IT Service Continuity Management Process Maturity |
| N | Number of Services With Test Failures |
| O | Total Number Of Services Needed To Support ITSCM Plans |
| P | Number Of Required Internal Support Services Without An OLA |
| Q | Number Of Required External Support Services Without Formal Agreements |

Suggested sources for IT Service Continuity Management Operational Metrics can be found in places such as:

- ✓ IT Service Continuity Plans
- ✓ Service Catalog Listings
- ✓ Service Continuity Test Plans
- ✓ Service Continuity Test Results
- ✓ HR and Labor Reports
- ✓ IT Financial Accounting and Budget Reports
- ✓ Operational Level Agreements (OLAs)
- ✓ Underpinning Contracts (UCs)
- ✓ Tool And Process Assessment Audit Findings

## *13.2 Key Performance Indicators (KPIs)*

The following table lists suggested KPIs and how they are calculated from the Operational Metrics listed previously.

| XREF | KPI | CALCULATION |
|---|---|---|
| 1 | ITSCM Coverage Ratio | *B/A* |
| 2 | Number of Service Continuity Plan Audit Failures | C |
| 3 | Mean Time (Months) Between Continuity Tests For Each Service | *D* |
| 4 | ITSCM Labor Utilization | *F/E* |
| 5 | Testing Completeness Ratio | *G/A* |
| 6 | ITSCM Cost Rate | *J/I* |
| 7 | IT Service Continuity Tooling Support Level | *L* |
| 8 | IT Service Continuity Management Process Maturity | *M* |
| 9 | IT Service Continuity Recovery Confidence Rate | *1-(N/A)* |
| 10 | Number of Business Continuity Issues Logged Against ITSCM Plans | *K* |
| 11 | Mean Time Between Continuity Plan Audits For Each Service | *H* |
| 12 | ITSCM Support Service Coverage Ratio | *1-((P+Q)/O)* |

## *13.3 Why These Metrics (KPIs) Matter*

The KPIs described earlier are critical to managing and controlling IT Service Continuity Management activities. The following table lists each IT Service Continuity Management KPI and the question it is trying to answer:

| KPI | Question Being Answered |
|---|---|
| ITSCM Coverage Ratio | How much of our IT Services are covered under the ITSCM Plan? |
| Number of Service Continuity Plan Audit Failures | How reliable is our ITSCM Plan? |
| Mean Time (Months) Between Continuity Tests For Each Service | How confident are we that enough testing has been done? |
| ITSCM Labor Utilization | How much of our planned labor capacity was used for ITSCM activities? |
| Testing Completeness Ratio | How complete is our testing? |
| ITSCM Cost Rate | How well did we plan for ITSCM Costs? |
| IT Service Continuity Tooling Support Level | How well does our current tool set support ITSCM activities? |
| IT Service Continuity Management Process Maturity | How well do we execute our ITSCM practices? |
| IT Service Continuity Recovery Confidence Rate | How much confidence do we have that we can recover needed services? |

| KPI | Question Being Answered |
|---|---|
| Number of Business Continuity Issues Logged Against ITSCM Plans | How aligned is our ITSCM plan with the Business Continuity Plan? |
| Mean Time Between Continuity Plan Audits For Each Service | How aligned is our ITSCM plan with our current infrastructure? |
| ITSCM Support Service Coverage Ratio | Do we have all necessary agreements with recovery support suppliers? |

## *13.4 Critical Success Factors (CSFs)*

The table below lists suggested Critical Success Factors for IT Service Continuity Management. The KPI references listed in the right column indicate which KPIs are used as input for the associated CSF.

| CSF | KPI |
|---|---|
| Recover Services From Major Disruptions Within Business Timeframes | *2,5,8,9,10,12* |
| Ensure All Required Services Can Be Recovered From Major Disruptions | *1,8,12* |
| Maintain Viability Of IT Service Continuity Plans | *3,8,10,11* |
| Provide Service Continuity At Acceptable Costs | *4,6,7* |

# Chapter 14

# IT Financial Management Metrics

## *14.1 Operational Metrics*

The following table lists suggested IT Financial Management Operational Metrics.

| XREF | METRIC |
|---|---|
| A | Total Infrastructure Budget |
| B | Total To-Date Planned Budget Costs |
| C | Total To-Date Actual Budget Costs |
| D | Financial Management Tooling Support Level |
| E | Financial Management Process Maturity |
| F | Total Actual Chargeback Revenue |
| G | Total Planned Chargeback Revenue |
| H | Number Of Financial Reports/Forecasts Delivered Late |
| I | Number Of Changes To Charging Algorithms |

Suggested sources for IT Financial Management Operational Metrics can be found in places such as:

- ✓ Chargeback Reports
- ✓ Capacity Plans
- ✓ IT Financial Accounting and Budget Reports
- ✓ Financial Reporting Production Schedules
- ✓ Tool And Process Assessment Audit Findings

## *14.2 Key Performance Indicators (KPIs)*

The following table lists suggested KPIs and how they are calculated from the Operational Metrics listed previously.

| XREF | KPI | CALCULATION |
|---|---|---|
| 1 | Unplanned Cost Variance | *B-A* |
| 2 | Cost Performance Index | *B/C* |
| 3 | Estimated Year-End Costs | *A/(B/C)* |
| 4 | Variance At Budget Year-End | *(A/(B/C))-A* |
| 5 | Financial Management Tooling Support Level | *D* |
| 6 | Financial Management Process Maturity | *E* |
| 7 | IT Revenue Variance | *H-G* |
| 8 | Number Of Financial Reports/Forecasts Delivered Late | *H* |
| 9 | Number Of Changes To Charging Algorithms | *I* |

## *14.3 Why These Metrics (KPIs) Matter*

The KPIs described earlier are critical to managing and controlling IT Financial Management activities. The following table lists each IT Financial Management KPI and the question it is trying to answer:

| KPI | Question Being Answered |
| --- | --- |
| Unplanned Cost Variance | Are we operating within planned costs? |
| Cost Performance Index | What is the ratio of expected costs to actual costs? |
| Estimated Year-End Costs | How much do we think we will actually spend by end of year? |
| Variance At Budget Year-End | How much more cost do we anticipate is needed by year-end? |
| Financial Management Tooling Support Level | How well does our current tool set support Financial Management activities? |
| Financial Management Process Maturity | How well do we execute our Financial Management practices? |
| IT Revenue Variance | How much chargeback revenue did we make compared to our plan? |
| Number Of Financial Reports/Forecasts Delivered Late | Are we delivering financial information to the business on time? |
| Number Of Changes To Charging Algorithms | Is there satisfaction with how IT charges for its services? |

## *14.4 Critical Success Factors (CSFs)*

The table below lists suggested Critical Success Factors for IT Financial Management. The KPI references listed in the right column indicate which KPIs are used as input for the associated CSF.

| CSF | KPI |
|---|---|
| Provide Effective Stewardship Of IT Finances | *1,2,3,4,6,8* |
| Maintain Overall Effectiveness Of The IT Financial Management Process | *5,6,8* |
| Ensure Customers Satisfied With Costs And Charges For Services | *9* |
| Recapture IT Costs Through Chargeback For Delivery Of It Services | *5,7* |

# Chapter 15

# IT Workforce Management Metrics

## *15.1 Operational Metrics*

The following table lists suggested IT Workforce Management Operational Metrics.

| XREF | METRIC |
|---|---|
| A | Total Number Of IT Service Delivery And Support Staff |
| B | Number Of IT Staff With Certifications In IT Service Management |
| C | Number Of IT Staff Actively Participating In Industry Quality Organizations |
| D | IT Employee Satisfaction Survey Rating |
| E | IT Staff Turnover Rate |
| F | Non-Value Labor Rating |

Suggested sources for IT Workforce Management Operational Metrics can be found in places such as:

- ✓ Labor and HR Reports
- ✓ Performance Reviews and Appraisals
- ✓ Employee Satisfaction Survey Results
- ✓ Employee Time Reporting Reports
- ✓ Employee Time Usage Surveys or Observations

## *15.2 Key Performance Indicators (KPIs)*

The following table lists suggested KPIs and how they are calculated from the Operational Metrics listed previously.

| XREF | KPI | CALCULATION |
|---|---|---|
| 1 | IT Staff Service Management Certification Rate | *B/A* |
| 2 | IT Staff Participation In External Industry Quality Organizations | *C/A* |
| 3 | IT Employee Satisfaction Survey Rating | *D* |
| 4 | IT Staff Turnover Rate | *E* |
| 5 | Non-Value Labor Rating | *F* |

## *15.3 Why These Metrics (KPIs) Matter*

The KPIs described earlier are critical to managing and controlling IT Workforce Management activities. The following table lists each IT Workforce Management KPI and the question it is trying to answer:

| KPI | Question Being Answered |
|---|---|
| IT Staff Service Management Certification Rate | How skilled is our IT staff in IT Service Management? |
| IT Staff Participation In External Industry Quality Organizations | How much of our IT staff participate in outside industry quality organizations? |
| IT Employee Satisfaction Survey Rating | How satisfied is our IT Staff with the work they do? |
| IT Staff Turnover Rate | How well do we retain IT staff, skills and experience? |
| Non-Value Labor Rating | How much IT labor is spent on non-value activities? |

## 15.4 Critical Success Factors (CSFs)

The table below lists suggested Critical Success Factors for IT Workforce Management. The KPI references listed in the right column indicate which KPIs are used as input for the associated CSF.

| CSF | KPI |
|---|---|
| Maximize IT Staff Productivity | 5 |
| Provide High Level Of Staff Skills In IT Service Quality | 1,2 |
| Maintain Positive Staff Morale | 3,4 |

# Chapter 16

# Alternatives If Few Metrics Available

## *16.1 Establishing a Minimal Metrics Program*

As efforts begin to establish an ITSM Metrics Program, there may be discovery that sources for metrics are far and few between. Typical challenges may include:

- Many disparate data collection and reporting tools in place with poor ability to aggregate and summarize data

- No clear authoritative source for metrics

- No staffing priorities to collect analyze and report on metrics

- A lack of tools and automation to report on metrics

When faced with these challenges, there is a great temptation to just quit the effort and wait until management priorities change or tool funding occurs. This is a great mistake. Attempts should always be made to measure and manage with what you can whenever possible. The alternative is to do no measuring at all.

Not measuring at all is a much worse alternative than selecting and agreeing on some minimal measurements that can be used as a starting point. These measurements may not be as accurate as those supported by a fully funded tool and project effort, but they can still serve until such an effort can get started.

The main benefit of establishing a metrics program with minimal metrics is to establish a culture of measurement goals and focus on IT service quality. Management desire to obtain better accuracy and information in this environment can actually lead to Senior Executive funding and approval for what will really be needed.

A general approach for establishing a set of minimal metrics can be done as follows:

1) Select a small subset of measurements that are "representative" of the quality of service being delivered

2) Develop assumptions as to their accuracy level and how they will be used

3) Review and agree these with senior management

4) Report on these as if a full-fledged metrics program were in place

Some examples of the kinds of minimal metrics that could be collected in this kind of program include the following:

- Use Indicators
- Random or Scheduled Inspection Results
- Analogous Measures
- Programmed Measures
- Audit Results

## *16.2 Use Indicators*

These are metrics that are based on some observable operational event for which an operating quality assumption is derived. Some examples might include:

- ✓ If the number of staff working on incidents throughout the day exceeds N, then it will be assumed that the acceptable service incident rate is too high
- ✓ If more than N incidents occur due to changes, then the Change Success Rate will be considered too low
- ✓ If the length of a service outage is N minutes, then the overall availability of that service will be assumed as

  *100% - (N/All planned availability minutes)*

## *16.3 Random or Scheduled Inspection Results*

These are metrics that represent observable events for which an assumption is made *that they apply to all similar events.* Some examples might include:

- ✓ If the utilization of network lines A, B and C are checked at 1:00pm on business days and found to be under N%, then it will be assumed that there is appropriate resource capacity to support services.
- ✓ If observed service restore times for the first 3 incidents of the week are X, Y and Z minutes, then it will be assumed that the Average Incident Restore time for all incidents is the average of those observations.
- ✓ If one remote network device has incurred N service outages it will be assumed that all remote network devices typically incur N service outages.
- ✓ Customer Satisfaction Survey results for one geographic location will represent similar results at all geographic locations.

## *16.4 Analogous Measures*

These are observable metrics from which assumptions will be derived that other events have occurred. In general, this approach states something like: "...if we see *this* – it really means *that* is happening..." This is purely a management decision and set of assumptions as to what makes an Analogous Measure. Some examples might include:

- ✓ If the number of customer complaint calls exceeds N on any given day of the month, then it will be assumed that the Customer Satisfaction level is too low.

- ✓ If a remote network device has incurred a service outage for N minutes then it will be assumed that all IT services at that location were also out of service for N minutes.

- ✓ If it is determined that an Incident was caused by a change and there is no RFC (Request For Change) that represents the Change, then the Number of Unauthorized Changes will increase by 1.

- ✓ The number of calls to the Service Desk will represent the number of incidents that took place.

- ✓ If the IT headcount at one operational facility is greater than 20% of the average headcount across all facilities, then it will be assumed that there is a high level of non-value labor at that facility.

## 16.5 Programmed Measures

Creating Programmed Measures involves developing programs that will produce events or measures that will substitute for real events. It is a management decision as to whether the programs developed will produce accurate findings or create random or analogous measures. That decision will usually be based on how much time, labor and cost that the business is willing to invest to produce the measure. Some examples might include:

- ✓ Developing end-to-end response time observations by creating dummy online transactions that traverse the network similar to real transactions. The dummy transactions drop a log file time stamp when they kick off and return. The observed response time is the difference between the two time stamps.

- ✓ Using the same program and approach as described above, the dummy transaction will be scheduled to execute every N minutes. If no response is received within Y minutes, it will be assumed that the IT service has experienced an outage and the Y value will be subtracted from the overall availability measure for the IT service it represents.

- ✓ An online transaction is programmed to update a special log file with the count of key business activities selected by a user that will be used to report on Business Capacity Management drivers.

## *16.6 Audit Results*

This consists of obtaining measures from periodic audit activities that are conducted for specific operational events. The results of the audit will then be applied to one or more KPI values. Some examples might include:

- ✓ Conducting a periodic Customer Satisfaction Survey and reporting the result as the Customer Satisfaction Level.
- ✓ Using a security penetration test to represent the quality of the availability of the IT infrastructure.
- ✓ Using an IT SOX or COBIT audit to assess the quality of the IT Change Management process.

## 16.7 Considerations for Using Minimal Metrics

The decision to use a Minimal Metrics Program should be based on the situation where metrics are really hard to obtain without a major project effort or tools investment. It should not be used as a quick shortcut substitute approach to avoid efforts to put in a quality Metrics solution.

Undertaking a minimal approach through the techniques described in this chapter will require a lot of creativity and management communications as to the measurement techniques being deployed. It means first identifying what the desired measurement set should be then identifying a set of techniques that will represent that measurement set.

At all times, the decision to undertake this kind of an effort will require constant management communications and a clear understanding and agreement to the techniques employed.

# Chapter 17

# Using the DICE Model

## *17.1 DICE Model Overview*

The DICE Model is a simple spreadsheet tool that can be used to predict the likelihood of success for an ITIL implementation or improvement effort. This model was originally developed by Harold L. Sirkin, Perry Keenan and Alan Jackson of the Boston Consulting Group who conducted a correlation study of 225 companies to determine the common denominators for successful organizational behavior change.

The DICE acronym represents the common denominators that were found. These are:

- **D**uration – length of time between project reviews
- **I**ntegrity – Extent to which the organization can rely upon the project team to execute the project successfully

- **C**ommitment – Ensuring that appropriate levels of Senior Management and Stakeholder commitment are in place

- **E**ffort – The estimated amount of time those making the change will have to spend over and above their day-to-day jobs

These four elements are then combined into a Project Success Prediction Score. In their study, regression analysis revealed that the combination of the above listed factors that correlated the most closely with actual project outcomes doubled the weight that was given to the performance of the team and the commitment of Senior Management. In the Model, the DICE Score Total is calculated from these factors using their formula.

For more specific information, readers are encouraged to read the white paper published by the authors. It can be found at:

Harvard Business Review: www.hbr.org
The Hard Side of Change Management
Harold L. Sirkin, Perry Keenan, Alan Jackson
Reprint: R0510G

## *17.2 Installing the Model*

The model is built as a Microsoft EXCEL Spreadsheet and included with the CD that accompanies this book. Simply download or copy this file to a desired folder on your PC. The PC itself should be running WINDOWS XP or other platform compatible with Microsoft Office 2003.

It is recommended that you install the original version of the file and make changes only to copies of it. This will allow you to continually reuse the original to create baselines or future state models of your project environment. For example:

1. Download and copy the original file to your PC as **DICE Model.xls**

2. Create a baseline of your project current state environment by making a copy of this file, applying your metrics results and storing it as **MyProject Baseline.xls** (for example)

3. Create state models of future project improvement decisions (such as modeling the impact of getting increased Target Stakeholder buy-in or a more capable Project Manager) by creating copies of your baseline model and storing it with some relevant name (i.e. **MyProjectImprovements.xls** for example) and then apply changes to that baseline. In this way, you can create multiple versions of models based on different project improvement scenarios and compare their impact to the overall DICE score results.

## *17.3 How to Use the DICE Model*

The model is simple to use. It consists of a single EXCEL Worksheet where you can input your project parameters. These are in the blue colored boxes. The remainder of the model, including the DICE prediction score and interpretation is automatically calculated for you.

## *17.4 Interpreting the Model Results*

The model will calculate a DICE Project Prediction score and then interpret this into the likelihood of success for your project into one of four areas. These are:

*Win Zone*
The project has a high likelihood that it will succeed.

*Worry Zone*
The project has a reasonable likelihood that it will succeed however there are some risk areas that should be watched closely. The risk areas are shown underneath the individual DICE Scores and color coded as Green, Yellow or Red. Any item coded as Red or Yellow with a score of 3 is a candidate area to watch closely.

*Woe Zone*
The project may not succeed unless certain risk areas are addressed. These are highlighted in Yellow or Red colors in the individual DICE Scores section of the model.

*Disaster Zone*
The project has a strong likelihood of NOT succeeding. The risk areas are highlighted in the individual DICE Scores section as described earlier.

## *17.5 Projects with Low Likelihood of Success*

The purpose of using the model is to highlight project risk areas that need to be addressed. If the calculated DICE score puts the project in the Woe or Disaster Zone, all is not lost. The project team simply needs to address the risks that are indicated in Red. An approach for doing this could be done as follows:

1) Identify which risk areas have the highest scores

2) Model changes to those risks. For example: If the capabilities of the project team were initially scored as LOW, change this to HIGH and see the overall impact to the project. Did it the resulting Zone change for the better?

3) Determine which changes had the greatest positive impact to your project effort

4) Develop an action plan to achieve the changes that you modeled. For example: add more highly skilled staff to your project team (from the previous example).

5) Obtain approval for your recommended changes and implement them

It is highly recommended that you re-model your project at each formal project review. This will allow you to identify risks early and mitigate them before they endanger your efforts. It may not be unusual to find that you may have started the effort with high stakeholder support initially, but that this has waned somewhat as the project proceeds over time. It is important to understand the impacts this could be having on your overall success.

# Chapter 18

# Using the ITSM Metrics Model Tool

## *18.1 ITSM Metrics Model Tool Overview*

The ITSM Metrics Model is a simple spreadsheet tool that can be used for a variety of measurement and reporting purposes. The model can be used as:

- ✓ A starting point to identify key metrics that can be used to measure and monitor the health and state of your ITSM processes and activities
- ✓ Justifying an ITSM improvement initiative by modeling desired target future state improvements expected to occur
- ✓ A means for demonstrating the impacts and effects of current ITSM practices

- ✓ A means for modeling future business decisions to assess their impact and risk to ITSM activities if those decisions were to take place

- ✓ A means for modeling the breaking point at which the quality of ITSM practices becomes untenable.

In short, this tool may be used to support ITSM reporting and to model the impact of changes to the IT infrastructure or future business decisions.

## *18.2 Installing the Model*

The model is built as a Microsoft EXCEL Spreadsheet and is included on the CD with this book. Simply download or copy this file to a desired folder on your PC. The PC itself should be running WINDOWS XP or other platform compatible with Microsoft Office 2003.

It is recommended that you install the original version of the file and make changes only to copies of it. This will allow you to continually reuse the original to create baselines or future state models of your ITSM environment. For example:

1. Download and copy the original file to your PC as **ITSM Metrics Model.xls**

2. Create a baseline of your ITSM current state environment by making a copy of this file, applying your metrics results and storing it as **ServiceBaseline.xls** (for example)

3. Create state models of your future ITSM and business decisions by creating copies of your baseline model and storing it with some relevant name (i.e. **PostMerger.xls** for example) and then apply changes to that baseline. In this way, you can create multiple versions of models based on different business scenarios and compare their risks and impacts. There will be more on this later.

When you first use the model, the values that are in the Operational and Tolerances sections (colored Yellow) are arbitrary. These have been put in there for placeholders. You will be replacing these with your real results.

## *18.3 How to Use the ITSM Metrics Model Tool*

The model is simple to use. It consists of an EXCEL Workbook with individual Worksheets for each ITSM process plus Service Desk and Workforce Worksheets. A Dashboard Worksheet is also included that averages and summarizes the results of all the ITSM processes. For each Worksheet, the steps are:

1) Fill out the **Tolerances** section of each worksheet with your **Target** and **Warning** values for each Key Performance Indicator (KPI)

2) Fill out the **Operational Metrics** section of each worksheet with values from ITSM tools, reports and observations

3) Optionally, determine whether you want to skip a particular Worksheet by changing the **Activate This Model** box from **Yes** to **No**. If you make this change, the model will assume that you ARE NOT performing this process and will RAISE modeled risk levels accordingly

4) Optionally, determine whether you want to exclude the results of a particular Worksheet by changing the **Add Results to Dashboard** box from **Yes** to **No**. It is not recommended that this option be taken since it means the Dashboard results will no longer reflect ALL the ITSM processes. Use this option only if you insist. It is there only for some organizations that want to see certain processes or combinations of processes without taking a holistic view of ITSM.

The model will automatically calculate the KPI values, compare them to the Tolerances and derive a LOW, MEDIUM or HIGH score with corresponding color (Green, Yellow, or Red) to indicate target status. A Green color indicates the KPI is at target or better. A Yellow color indicates the KPI is between the target and warning levels. A Red color indicates the target is above or below the warning level.

Critical Success Factors (CSFs) are automatically calculated based on the KPI values. A CSF consists of one or more KPIs that relate to it. It is color coded based on how well the combination of its KPIs were averaged. Therefore, a Red color indicates the CSF is at high risk, Yellow at Medium risk or Green at Low risk.

CSFs are then factored into the Worksheet level dashboard section. The balanced scorecard boxes (Customer, Capability, Operational, Financial and Regulatory) are derived from one or more combinations of CSFs that impact them. The same occurs with each of the operational risk areas (Legal Exposure, Service Outages, Rework, etc.). For these, the dashboard colors indicate the possible likelihood that exists for each risk occurring.

The Dashboard Worksheet is then calculated automatically as a rollup from each of the individual process dashboards.

The data flows within the model are fairly simple. It can best be described using the metrics model previously presented:

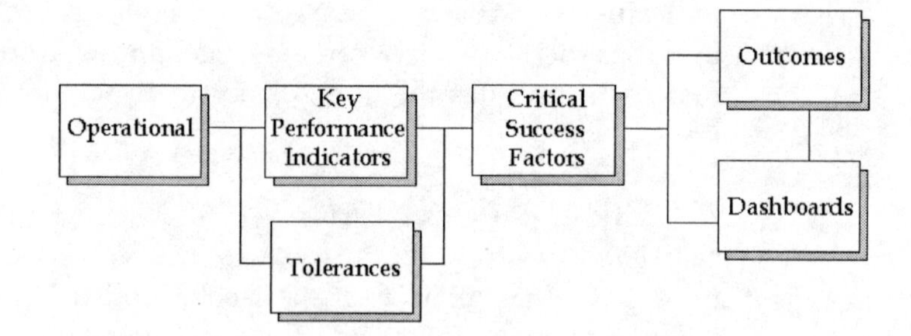


1) Tolerances are first entered for each process to describe acceptable and not acceptable KPI levels

2) Operational metrics are then entered for each process with live data from ITSM process reporting and other infrastructure measurements and observations

3) Key Performance Indicators (KPIs) are then calculated from the above and coded Green, Yellow or Red depending on how they fell within the specified Tolerance Levels

4) CSF risk levels are then calculated from combinations of KPI results and color coded as Green (Low), Yellow (Medium) or Red (High)

5) Each individual process Dashboard is then calculated from combinations of CSF results

6) The dashboard worksheet is calculated from averages across all CSFs and Dashboard risk results.

## *18.4 Interpreting the Model Results*

There are four items of interest that are output from this tool:

- ✓ KPI results
- ✓ CSF Results
- ✓ Balanced Scorecard Results
- ✓ Risk Assessment Results

<u>*KPI Results*</u>

These are the "Metrics That Matter". An example from the Change Management worksheet in the model is shown below:

| Key Performance Indicators (KPIs) | | *Question To Be Answered* |
|---|---|---|
| Change Efficiency Rate | 95.0% | How efficient are we at handling changes? |
| Change Success Rate | 97.4% | How effective are we at handling changes? |
| Emergency Change Rate | 2.5% | What percentage of changes were emergencies? |
| Change Reschedule Rate | 5.0% | How well do we implement changes on schedule? |
| Average Process Time Per Change (Days) | 2.9 | How long does the average change take? |
| Unauthorized Change Rate | 2.1% | What percentage of changes bypassed the Change process? |
| Change Incident Rate | 0.3% | What percentage of changes caused incidents? |
| Change Labor Workforce Utilization | 78.0% | How much available labor capacity was spent handling changes? |
| Change Management Tooling Support Level | 2.4 | How well does our current tool set support Change Management activities? |
| Change Management Process Maturity | 2.4 | How good is our Change Management practices? |

The results for each KPI are shown as calculated from the Operational Metrics input earlier. Color coding is based on how well the KPI fell within the Target and Warning Tolerance levels. Red results indicate potential areas that need to be improved.

As an example, the *Change Success Rate* KPI is calculated as follows from the Operational Metrics:

*Number of Failed Changes /Total Changes Implemented*

Therefore, if you implemented 1,000 changes and had no failures you scored a 100%. If you had 100 changes fail, you scored a 90%. This result is compared against what you input for Tolerance target and warning levels. If you indicated that your target was 98% and the warning level was 85% (as an example), then a 100% score would appear green. The 90% score would appear yellow. If your score ended up as, 84% for example, the score would appear red.

<u>*CSF Results*</u>

These are derived from specific KPIs that relate to them. An example from the Change Management worksheet in the model is shown below:

| Critical Success Factors | *Target Level* |
|---|---|
| Protect Services When Making Changes | High |
| Make Changes Quickly And Accurately In Line With Business Needs | Medium |
| Make Changes Efficiently And Effectively | High |
| Utilize A Repeatable Process For Handling Changes | Medium |

In the above, "Protect Services When Making Changes" is calculated from the following KPIs:

- ✓ Emergency Change Rate
- ✓ Unauthorized Change Rate
- ✓ Change Incident Rate

These KPIs were chosen because they relate to specific threats to "Protecting Services When Making Changes". The model examines each of those KPIs and then provides a result equal to the KPI with the highest risk. Therefore, for example, this result could score a Red (High) in the situation where a low Emergency and Unauthorized Change rate existed with a high Change Incident Rate.

*Balanced Scorecard Results*

A high level Balanced Scorecard is presented that looks like the following:

| Customer | | Capability | | Operational | | Financial | | Regulatory |
|---|---|---|---|---|---|---|---|---|
| Medium | | Medium | | Medium | | High | | None |
| | | | | | | | | |
| 4.0 | | 9.0 | | 6.0 | | 3.0 | | 0.0 |
| 2 | | 4 | | 3 | | 2 | | 1 |

This is simply a showing of each scorecard area (Customer, Capability, Operational, Financial and Regulatory) with a color coded result that indicates the risk level for each area (Green (low), Yellow (medium), and Red (high).

The first row of numbers underneath represents a calculated score total. This is taken from the mean average of all the CSFs that were input to that score. The second row simply indicates the number of CSFs that were used to calculate the score.

In the above example, the Capability Score is 9.0 and there were 4 CSFs that fell into that area. These values are used to determine the High, Medium or Low rating for the scorecard box. They are there only for calculation purposes.

*Risk Assessment Results*

Risk assessment results are also included with each Dashboard. These represent outcomes derived from the KPIs and CSFs. They are pictured as follows:

| | | | | |
|---|---|---|---|---|
| Low | ➡ | Legal Exposure | 1.0 | 1 |
| Low | ➡ | Service Outages | 1.0 | 1 |
| Medium | ➡ | Rework | 4.0 | 2 |
| Medium | ➡ | Waste | 2.0 | 1 |
| High | ➡ | Delayed Solutions | 3.0 | 1 |
| Medium | ➡ | Slow Operational Processes | 2.0 | 1 |
| None | ➡ | Security Breaches | 0.0 | 1 |
| None | ➡ | Inaccurate Information | 0.0 | 1 |
| High | ➡ | Slow Turnaround Times | 3.0 | 1 |
| Medium | ➡ | Unexpected Costs | 9.0 | 4 |
| Medium | ➡ | Higher or escalating costs | 8.0 | 3 |
| Low | ➡ | Low Employee Morale | 1.0 | 1 |
| High | ➡ | Slow Response To Business Needs And Changes | 3.0 | 1 |
| Low | ➡ | Unwanted PR Exposure | 1.0 | 1 |
| Medium | ➡ | Dissatisfied Customers | 4.0 | 2 |
| None | ➡ | Dissatisfied Suppliers | 0.0 | 1 |
| Medium | ➡ | Inability to scale | 5.0 | 2 |
| Medium | ➡ | Fines and Penalties | 4.0 | 2 |
| Low | ➡ | High Levels Of Non-Value Labor | 3.0 | 2 |
| Medium | ➡ | Loss of Market Share | 4.0 | 2 |
| Medium | ➡ | Loss of Revenue/Sales | 6.0 | 3 |

Each risk area represents a possible outcome based on how well CSFs were met. The color coded box to the left of each outcome indicates the likelihood that the associated risk might occur.

The number values to the right of each risk are similar to the dashboard results except they are specific to each risk. These are used to calculate the High, Medium, Low rating and color for each risk identified.

## *18.5 Modeling Business Decisions*

One of the main purposes of the tool is to model the impact of business decisions or ITSM improvements that you are thinking about or planning to make. Examples of questions you may be trying to answer might be as follows:

- ✓ What will be the impact on our IT service quality if we put a major new application into production?
- ✓ How much operational risk will occur if we go through with a planned merger or acquisition?
- ✓ Which ITSM improvement initiatives will provide us with the most benefit?
- ✓ How many problems, incidents or changes can we handle before the quality of our services breaks down?
- ✓ What is the impact of increasing our Change Management CMMI Process Maturity from 2.4 to 3.5?

If using the tool to model things like this, you will first create a Baseline ITSM Model. The Baseline will represent your Current State practices. This model will be populated with the results of the way you currently utilize ITSM processes and activities. In this, you will populate the Tolerances and Operational Metrics as described earlier.

The next step is to create a series of Future State ITSM Models. These are also known as "What-If" Models. For each of these models you will make a replica of your baseline model and then make various changes to it that reflect various scenarios that you would like to model.

As an example, let's say that most of your ITSM Processes are at a 2.0 level. What might be the impact if you raise this level to 3.0? For this, you would:

1) Make a replica of your baseline ITSM model

2) Change the Process Maturity to 3.0 in the Operational Metrics section of each process you are interested in

3) View the results

You could then save this model for future reference and build other models to reflect other scenarios such as:

- ✓ What happens if the volumes of Incidents are decreased by 20%?

- ✓ What happens if the Emergency Change rate rises by 30%?

- ✓ What happens if Change Management labor is decreased by 10%?

Remember that only the Operational Metrics will be changed for What-If models. The model will calculate whether those changes resulted in KPIs that fell out of (or into) desired Tolerance levels.

# Chapter 19

# Implementing an ITSM Metrics Program

## *19.1 ITSM Metrics Program End State*

The ITSM Metrics Program is a key effort towards supporting the ITSM continuous improvement cycle. Remember, what can't be reported on can't be improved. A suggested vision for an ongoing ITSM Metrics Program will include actions that take place on a scheduled periodic basis to:

- ✓ Align metrics to current business need
- ✓ Report on CSF and KPI results
- ✓ Review and analyze those results
- ✓ Identify results that fall below performance goals
- ✓ Initiate actions to bring failing results back to acceptable performance levels

It is highly recommended that these actions take place monthly if at all possible, but no less frequently than once every business quarter.

Key roles that will be needed to execute the metrics program include the following:

**Metrics Program Leader**

This role will be responsible for the overall ITSM Metrics Program. It ensures metrics are collected, produced and reported on within acceptable timeframes. It acts as a single point of contact to executive management for program results and actions being taken.

**Process Owner**

Per ITIL guidelines, the Process Owner role is responsible for one or more ITIL processes. It's mission is to ensure the process is running efficiently, cost effectively and meeting business goals. In the context of the ITSM Metrics Program, this role is also responsible for providing the needed Operational and Tolerance metrics and identifying and initiating improvement actions if those metrics fall below Tolerance levels.

**Metrics Analyst**

This role is responsible for collecting, aggregating, summarizing and reporting on metrics data. It coordinates all activities needed to produce the metrics reports. It may also coordinate activities to produce one-time drill-downs or other views of data needed to support an improvement analysis effort. In addition, this role also maintains a repository of historical metrics information and reports.

**Program Sponsor**

This role covers the key senior executive who has approved and funded the overall metrics program.

**Program Stakeholder**

This role represents the customer of the metrics program. The stakeholder will review the metric results, identify concerns and issues, and agree on improvement actions when needed. Should metrics fall below Tolerance levels, it is this role that will prioritize, agree and fund actions to make improvement efforts happen.

**Tool Architect**

This role maintains the overall program tooling solution architecture and ensures that tooling solutions integrate properly and align with business need. In addition, this role may also be consulted for improvements and enhancements to tooling solutions that support the program over time.

**Tool Technician**

This role is responsible for maintaining and supporting the current tooling infrastructure that supports the metrics program.

**Technical Writer**

This role is responsible for documenting Program training materials, process, procedure and architecture guides.

A brief synopsis of day-to-day life in the metrics program at each reporting period might look like the following:

*Align metrics to current business need*

In this task, the team gets together with metrics stakeholders and identifies any needed changes to the metrics program. Are the current metrics still sufficient? Are changes needed to CSFs and KPIs? Is the current quality of reporting efficient? Changes are identified and scheduled to be in place for future reporting periods.

*Report on CSF and KPI results*

In this task, the team assembles aggregates and summarizes metrics data and assembles them into reports and/or dashboards. The reports are then distributed or placed online where they can be accessed and viewed. Metric information and reports are also stored for historical purposes in a metrics repository.

*Review and analyze those results*

In this task, the Program Leader, Process Owners, and the Metrics Analyst get together and take an initial pass at the results to determine that they are fair, accurate and representative of what took place during the previous reporting period.

*Identify results that fall below performance goals*

The Process Owners check to see which KPIs and CSFs have fallen below acceptable performance levels (Tolerance levels). For each KPI, the Process Owners will develop action plans and alternatives to bring those results back to accepted levels. Approaches for doing this are then summarized for senior executive management.

*Initiate actions to bring failing results back to acceptable performance levels*

In this task, the Program Leader and Process Owners meet with metrics stakeholders and executive senior management to go over improvement action alternatives. Examples of outcomes of such a meeting might include:

- Agree and fund one or more projects to address needed improvements to correct the deficiencies found
- Adjust tolerance levels to reflect closer reality to what can actually be delivered and achieved
- Agree to monitor deficiencies for a set period of time and delay actions to make sure deficiencies are not one-time events

A summary of roles and responsibilities for the ITSM Metrics Program as suggested is presented below *(Note: the Technical Writer is not typically needed for operating the program, only building it)*:

| Key Task | Program Leader | Process Owner | Metrics Analyst | Program Sponsor | Stakeholder(s) | Tool Architect | Tool Technician |
|---|---|---|---|---|---|---|---|
| Align metrics to current business need | A | R | C | S | C | C | I |
| Report on CSF and KPI results | A | C | R | S | I | I | C |
| Review and analyze those results | R | A | C | S | I | I | I |
| Identify results that fall below performance goals | A | R | C | S | I | I | I |
| Initiate actions to bring failing results back to acceptable performance levels | R | A | I | C | S | C | C |

S = Signatory
A = Accountable
R = Responsible
C = Consults
I = Informed

The importance of having an ongoing ITSM Metrics Program cannot be understated. Without this, it is impossible to accurately identify needed service improvements on a timely basis.

Without this, IT cannot demonstrate that it can effectively govern itself and align its priorities with what is needed for the business.

## *19.2 Work Breakdown Structure*

The suggested Work Breakdown Structure that represents the key deliverables of an ITSM Metrics Program Implementation Effort can be shown as follows:

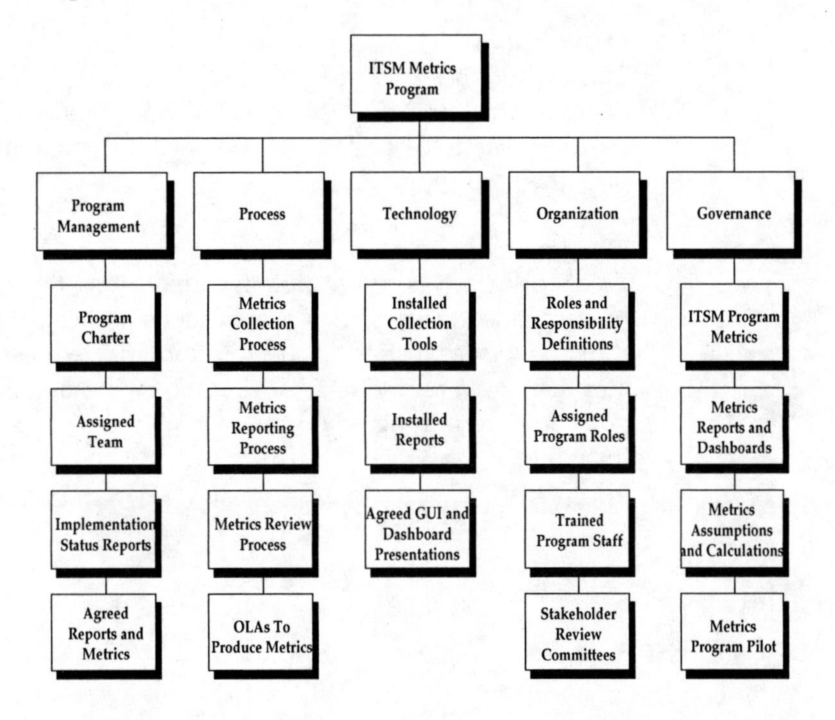


The above deliverables are divided into 5 project tracks that can work in parallel:

**Program Management**
Oversees and manages the implementation effort

**Process**
Develops processes needed to operate and run the program on an ongoing basis. This also includes establishing OLA (Operational Level Agreements) as needed to obtain metrics information or produce reports.

**Technology**
This work track is responsible for developing the program solution tooling architecture. It also is responsible for procuring, installing and testing of those tools.

**Organization**
This work track is responsible for identifying and documenting program roles and responsibilities. It also ensures that roles have been assigned to program staff once they are approved.

**Governance**
This work track is responsible for identifying the needed metrics, documenting their assumptions and calculations. It also oversees testing of the solution and will establish and manage a pilot effort. Governance will have the key voice in determining when the metrics program is ready to go live.

## *19.3 Implementation Approach*

The suggested work approach uses 5 key work steps to implement the ITSM Metrics Program:

1. Program Initiation
2. Program Design
3. Program Building and Testing
4. Program Pilot
5. Program Rollout and Transition

Ideally, these work steps should be incorporated into an overall ITSM ITIL Process Implementation Program as part of the Governance tasks for that effort. For the purposes of this book, we will assume that this program is being implemented standalone to better outline suggested tasks that need to be completed.

It is suggested that the Program be steered by and follow the ITIL Release and Change Management processes. A more detailed explanation of each step is as follows:

**Program Initiation**

This set of tasks involves developing the Program Charter and preliminary scope for the effort. The preliminary scope will outline what business units are covered by the effort, key CSFs and KPIs that are initially desired, who will receive and act on metrics reporting and any key assumptions. This effort ends with an approved program implementation project, assigned project lead and appropriate funding needed to complete its goals.

A list of tasks for this work step is as follows:

| **Initiation Work Tasks** | **Program Leader** | **Process Owner** | **Metrics Analyst** | **Program Sponsor** | **Stakeholder(s)** | **Tool Architect** | **Tool Technician** | **Technical Writer** |
|---|---|---|---|---|---|---|---|---|
| ***Program Management*** | | | | | | | | |
| Develop Preliminary Program Scope | A | C | I | C | C | I | I | I |
| Build ITSM Metrics Program Charter | A | C | I | C | C | I | I | I |
| Approve ITSM Metrics Program Charter | A | C | I | S | C | I | I | I |
| Assign Implementation Team | A | C | I | S | C | I | I | I |
| Build Program Work Plan | A | C | C | S | C | C | C | C |
| Conduct Program Kickoff Meeting | A | C | C | I | C | C | C | C |

*(S=Signatory, A=Accountable, R=Responsible, C=Consults, I=Informed)*

### Program Design

In this work step, design tasks take place to identify the metrics that will be used, definitions, key assumptions and calculations. Identification of how each metrics will be sourced, what reporting tools, report and dashboard templates will also take place. Processes for collection, and reporting will be designed and roles and responsibilities for these will also be identified.

A list of tasks for this work step is as follows:

| Design Tasks | Program Leader | Process Owner | Metrics Analyst | Program Sponsor | Stakeholder(s) | Tool Architect | Tool Technician | Technical Writer |
|---|---|---|---|---|---|---|---|---|
| ***Governance*** | | | | | | | | |
| Identify ITSM Program Metrics | A | C | R | S | C | C | I | I |
| Define Assumptions and Calculations | A | C | R | C | C | C | I | I |
| Design Metrics Reports and Dashboards | A | C | C | S | C | C | C | I |
| Agreed GUI and Dashboards | A | C | C | S | C | C | C | I |
| ***Organization*** | | | | | | | | |
| Define Program Roles and Responsibilities | A | C | C | S | C | C | I | I |
| Define Program Skills | A | C | C | S | C | C | I | I |
| Identify Program Training Requirements | A | C | C | S | C | C | C | I |
| Identify Stakeholder Review Committee | A | C | I | S | C | I | I | I |
| Develop Program Communications Plan | A | C | I | S | C | I | I | I |
| ***Process*** | | | | | | | | |
| Develop Metrics Collection | A | C | R | I | C | C | C | I |

| **Design Tasks** | **Program Leader** | **Process Owner** | **Metrics Analyst** | **Program Sponsor** | **Stakeholder(s)** | **Tool Architect** | **Tool Technician** | **Technical Writer** |
|---|---|---|---|---|---|---|---|---|
| Processes | | | | | | | | |
| Develop Metrics Reporting Process | A | C | R | I | C | C | C | I |
| Develop Review Committee Agenda | A | C | R | C | C | I | I | I |
| Develop Metrics Review Process | A | C | R | I | C | I | I | I |
| Draft Metrics Production Guide | A | C | C | S | C | C | C | R |
| ***Technology*** | | | | | | | | |
| Identify Metric Sourcing Strategy | A | C | C | I | C | R | C | I |
| Identify Metrics Repository Architecture | A | C | C | I | C | R | C | I |
| Identify Metrics Collection Tools | A | C | C | I | C | R | C | I |
| Identify Metrics Reporting Tools | A | C | C | I | C | R | C | I |
| Identify Metrics Presentation Tools | A | C | C | I | C | R | C | I |
| Drafty Metrics Architecture | A | C | C | S | C | R | C | C |
| Agreed GUI and Dashboards | A | C | C | S | C | R | C | I |
| ***Program Management*** | | | | | | | | |
| Update Implementation Plans | A | C | C | C | C | C | C | C |
| Integrate Project Activities· | A | I | I | C | C | I | I | I |
| Report Program Status | A | C | C | C | I | C | C | C |

### Program Building and Testing

In this work step, tasks are done to build and test the Program tools and architecture. This includes procuring and assembling needed tools and documenting detailed procedures for metrics collection and reporting. It also includes unit and integration testing of program tools and processes.

It is in this step that a Pilot be selected for the Program. The Pilot may consist of:

- ✓ One or more business units that will be covered under the metrics program for a set period of time
- ✓ Covering all business units in a "trial" mode for a set period of time
- ✓ Some combination of the above

The purpose of the Pilot is to ensure that the Program is operating as planned and that acceptance for it is gained. Therefore, expectations with Pilot candidates should be clearly set to let staff know that some errors and issues may occur during this time. Without this, a perception may develop that the Program is flawed. These kinds of communications need to be established in advance as part of this work effort.

A list of tasks for this work step is as follows:

| Build and Test Tasks | Program Leader | Process Owner | Metrics Analyst | Program Sponsor | Stakeholder(s) | Tool Architect | Tool Technician | Technical Writer |
|---|---|---|---|---|---|---|---|---|
| ***Governance*** | | | | | | | | |
| Develop Test Conditions and Results | A | C | R | C | C | C | C | I |
| Oversee and Review Test Results | A | C | R | C | C | C | C | I |
| Select and Agree Pilot | A | C | R | S | C | C | C | I |
| Draft Pilot Plan | A | C | C | C | C | C | C | I |
| Oversee Training For Pilot Staff | A | C | C | I | I | I | I | I |
| ***Organization*** | | | | | | | | |
| Update Job Descriptions For Roles | A | C | C | S | C | C | C | I |
| Create Program Training Materials | A | C | C | S | C | C | C | R |
| Assign Personnel For Pilot | A | C | C | S | C | C | C | I |
| Conduct Program Training For Pilot | A | C | C | I | I | C | C | C |
| ***Process*** | | | | | | | | |
| Develop Detailed Procedures | A | C | R | I | C | C | C | I |
| Test Processes and Procedures | A | C | R | I | C | C | C | I |
| Update Metrics Production Guide | A | C | C | C | C | C | C | R |
| ***Technology*** | | | | | | | | |
| Procure Program Tools | A | C | C | S | C | R | C | I |
| Install and Customize Tools | A | C | C | I | C | R | C | I |
| Test Program Architecture | A | C | C | I | C | R | C | I |
| Validate Program Reports/Presentations | A | C | C | S | C | R | C | I |
| ***Program Management*** | | | | | | | | |

| Build and Test Tasks | Program Leader | Process Owner | Metrics Analyst | Program Sponsor | Stakeholder(s) | Tool Architect | Tool Technician | Technical Writer |
|---|---|---|---|---|---|---|---|---|
| Update Implementation Plans | A | C | C | C | C | C | C | C |
| Integrate Project Activities | A | I | I | C | C | I | I | I |
| Report Program Status | A | C | C | C | I | C | C | C |

**Program Pilot**

In this work step, the Program Pilot is executed. This involves running and operating the Pilot as if the Program were running live in real production. The Pilot should be monitored for deficiencies. It is important that these are addressed right away when discovered or the effort will appear flawed. During this period, the Program team is executing two main tasks: supporting the Pilot and making preparations for final rollout of the Program solution.

A list of tasks for this work step is as follows:

| Pilot Tasks | Program Leader | Process Owner | Metrics Analyst | Program Sponsor | Stakeholder(s) | Tool Architect | Tool Technician | Technical Writer |
|---|---|---|---|---|---|---|---|---|
| ***Governance*** | | | | | | | | |
| Oversee and Manage Pilot Results | A | C | C | C | C | C | C | I |
| Maintain Inventory of Deficiencies | A | C | R | C | C | C | C | C |
| Ensure Deficiencies Addressed | A | C | C | C | C | C | C | C |
| Draft Program Rollout Plans | A | C | C | C | C | C | C | I |
| Agree Program Rollout Plans | A | C | C | S | C | C | C | I |
| ***Organization*** | | | | | | | | |
| Monitor Pilot For Skill Levels | A | C | C | I | C | C | C | I |
| Address Training Changes If Needed | A | C | C | C | C | C | C | C |
| Manage Pilot Stakeholders | A | C | I | C | C | C | I | I |
| ***Process*** | | | | | | | | |
| Monitor Pilot For Process Deficiencies | A | C | C | I | C | I | I | I |
| Address Process Deficiencies If Found | A | C | R | C | C | C | C | I |
| Support Pilot Processes | A | C | C | C | C | C | C | I |
| ***Technology*** | | | | | | | | |
| Monitor Pilot for Technical Deficiencies | A | C | C | I | C | R | C | I |
| Address Technical Issues If Found | A | I | I | I | I | R | C | I |
| Support Pilot Technical Architecture | A | I | I | I | I | R | C | I |
| ***Program Management*** | | | | | | | | |
| Update Implementation Plans | A | C | C | C | C | C | C | C |

| Pilot Tasks | Program Leader | Process Owner | Metrics Analyst | Program Sponsor | Stakeholder(s) | Tool Architect | Tool Technician | Technical Writer |
|---|---|---|---|---|---|---|---|---|
| Integrate Project Activities | A | I | I | C | C | I | I | I |
| Report Program Status | A | C | C | C | I | C | C | C |

## Program Rollout and Transition

In this work step, the ITSM Metrics Program is rolled out to the rest of the organization. This involves executing the rollout plans. Operational tasks for the program are converted to production and remaining additional staff (if needed) are trained and put into place.

At the end of this step, the ITSM Metrics Program should be fully operational. The only additional task at this point is to initiate ongoing review and monitoring of the Program to identify future improvement enhancements. The suggested approach described for implementation can be reused in abbreviated form as needed to implement improvements over time.

A list of tasks for this work step is as follows:

| **Rollout and Transition Tasks** | **Program Leader** | **Process Owner** | **Metrics Analyst** | **Program Sponsor** | **Stakeholder(s)** | **Tool Architect** | **Tool Technician** | **Technical Writer** |
|---|---|---|---|---|---|---|---|---|
| ***Governance*** | | | | | | | | |
| Oversee and Manage Program Rollout | A | C | C | S | C | C | C | I |
| Maintain Inventory of Deficiencies | A | C | R | C | C | C | C | C |
| Ensure Deficiencies Addressed | A | C | C | C | C | C | C | C |
| Validate Program Goals Achieved | A | C | C | S | C | C | C | I |
| Identify Future Improvements | A | C | C | C | C | C | C | I |
| ***Organization*** | | | | | | | | |
| Conduct Planned Training Activities | A | C | C | I | C | C | C | I |
| Identify Training Improvements | A | C | C | C | C | C | C | I |
| Manage Program Stakeholders | A | C | C | C | C | C | C | C |
| ***Process*** | | | | | | | | |
| Monitor Program For Deficiencies | A | C | C | I | C | I | I | I |
| Identify Process Improvements Needed | A | C | R | C | C | C | C | I |
| ***Technology*** | | | | | | | | |
| Support Program Technical Architecture | A | I | I | I | I | R | C | I |
| Identify Future Technical Improvements | A | I | I | I | I | R | C | I |
| ***Program Management*** | | | | | | | | |
| Report Program Status | A | C | C | C | I | C | C | C |
| Close Project | A | I | I | S | I | I | I | I |

# *About the Author*

**Randy A. Steinberg** has over 25 years experience implementing and managing IT infrastructures and service management solutions at numerous companies around the world. His past roles include a stint as Global Head of Service Management for Reuters, Head of Operations and Support for the Milwaukee Medical Complex and many Lead Manager roles for numerous large client infrastructure implementations for IBM and Accenture clients. Randy has been a frequent speaker at a number of itSMF meetings around the U.S. and was invited to keynote the national itSMF convention in 2004.

Randy is Manager Certified in IT Service Management. He is the co-author of an ITSM methodology and operational framework that was used worldwide by Andersen Consulting. One of his IT Service Management clients received a Malcolm Baldridge award for the quality of their IT services.

Randy is currently a Senior Consultant at Covestic where he leads large scale ITSM efforts for their clients. He is author of the book:

*Implementing ITIL – Adapting Your Organization to the Coming Revolution in IT Service Management*

*Trafford Press: ISBN 1-4120-6618-2*

Feel free to contact Randy about any ITSM related concerns, issues or recommended changes and additions to this book. They are always welcome. Randy can be reached at **RandyASteinberg@aol.com.**

ISBN 141209392-9
9 781412 093927